Tumors of the Serosal Membranes

AFIP Atlas
of
Tumor Pathology

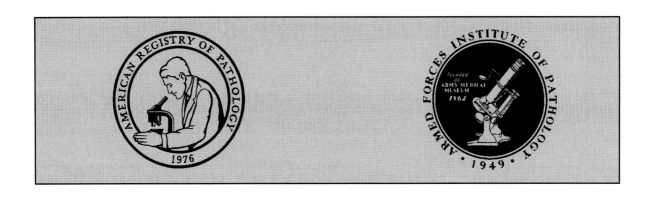

Available from the American Registry of Pathology
Armed Forces Institute of Pathology
Washington, DC 20306-6000
www.afip.org
ISBN 1-881041-97-2

ARP PRESS

Silver Spring, Maryland

Editorial Director: Kelley S. Hahn
Production Editor: Dian S. Thomas
Editorial/Scanning Assistant: Mirlinda Q. Caton
Copyeditor: Audrey Kahn
Scanning Technician: Kenneth Stringfellow

AFIP ATLAS OF TUMOR PATHOLOGY

Fourth Series
Fascicle 3

TUMORS OF THE SEROSAL MEMBRANES

by

ANDREW CHURG, MD
Professor of Pathology
University of British Columbia
Pathologist
Vancouver Hospital, Vancouver, Canada

PHILIP T. CAGLE, MD
Professor of Pathology
Baylor College of Medicine
Attending Pathologist
The Methodist Hospital, Houston, Texas

VICTOR L. ROGGLI, MD
Professor of Pathology
Duke University and Durham VA Medical Centers
Durham, North Carolina

Published by the
American Registry of Pathology
Washington, DC
in collaboration with the
Armed Forces Institute of Pathology
Washington, DC
2006

AFIP ATLAS OF TUMOR PATHOLOGY

EDITOR
Steven G. Silverberg, MD
Department of Pathology
University of Maryland School of Medicine
Baltimore, Maryland

ASSOCIATE EDITOR
Leslie H. Sobin, MD
Armed Forces Institute of Pathology
Washington, DC

EDITORIAL ADVISORY BOARD

Manuscript Reviewed by:
William J. Frable, MD
Nelson G. Ordonez, MD

EDITORS' NOTE

The Atlas of Tumor Pathology has a long and distinguished history. It was first conceived at a Cancer Research Meeting held in St. Louis in September 1947 as an attempt to standardize the nomenclature of neoplastic diseases. The first series was sponsored by the National Academy of Sciences-National Research Council. The organization of this Sisyphean effort was entrusted to the Subcommittee on Oncology of the Committee on Pathology, and Dr. Arthur Purdy Stout was the first editor-in-chief. Many of the illustrations were provided by the Medical Illustration Service of the Armed Forces Institute of Pathology (AFIP), the type was set by the Government Printing Office, and the final printing was done at the Armed Forces Institute of Pathology (hence the colloquial appellation "AFIP Fascicles"). The American Registry of Pathology (ARP) purchased the Fascicles from the Government Printing Office and sold them virtually at cost. Over a period of 20 years, approximately 15,000 copies each of nearly 40 Fascicles were produced. The worldwide impact of these publications over the years has largely surpassed the original goal. They quickly became among the most influential publications on tumor pathology, primarily because of their overall high quality but also because their low cost made them easily accessible the world over to pathologists and other students of oncology.

Upon completion of the first series, the National Academy of Sciences-National Research Council handed further pursuit of the project over to the newly created Universities Associated for Research and Education in Pathology (UAREP). A second series was started, generously supported by grants from the AFIP, the National Cancer Institute, and the American Cancer Society. Dr. Harlan I. Firminger became the editor-in-chief and was succeeded by Dr. William H. Hartmann. The second series' Fascicles were produced as bound volumes instead of loose leaflets. They featured a more comprehensive coverage of the subjects, to the extent that the Fascicles could no longer be regarded as "atlases" but rather as monographs describing and illustrating in detail the tumors and tumor-like conditions of the various organs and systems.

Once the second series was completed, with a success that matched that of the first, ARP, UAREP, and AFIP decided to embark on a third series. Dr. Juan Rosai was appointed as editor-in-chief, and Dr. Leslie H. Sobin became associate editor. A distinguished Editorial Advisory Board was also convened, and these outstanding pathologists and educators played a major role in the success of this series, the first publication of which appeared in 1991 and the last (number 32) in 2003.

The same organizational framework will apply to the current fourth series, but with UAREP no longer in existence, ARP will play the major role. New features will include a hardbound cover, illustrations almost exclusively in color, and an accompanying electronic version of each Fascicle. There will also be increased emphasis (wherever appropriate) on the cytopathologic (intraoperative, exfoliative, and/or fine needle aspiration) and molecular features that are important in diagnosis and prognosis. What will not change from the three previous series, however, is the goal of providing the practicing pathologist with thorough, concise, and up-to-date information on the nomenclature and classification; epidemiologic, clinical, and pathogenetic features; and, most importantly, guidance in the diagnosis of the tumors and tumorlike lesions of all major organ systems and body sites.

As in the third series, a continuous attempt will be made to correlate, whenever possible, the nomenclature used in the Fascicles with that proposed by the World Health Organization's Classification of Tumors, as well as to ensure a consistency of style throughout. Close cooperation between the various authors and their respective liaisons from the Editorial Board will continue to be emphasized in order to minimize unnecessary repetition and discrepancies in the text and illustrations.

Particular thanks are due to the members of the Editorial Advisory Board, the reviewers (at least two for each Fascicle), the editorial and production staff, and—first and foremost—the individual Fascicle authors for their ongoing efforts to ensure that this series is a worthy successor to the previous three.

Steven G. Silverberg, MD
Leslie H. Sobin, MD

PREFACE

The existence of tumors arising in the serosal membranes was for many years a subject of intense controversy; indeed, even in the mid to late 1960s some standard textbooks on tumor pathology claimed that there was no such thing as a malignant mesothelioma. It was really the demonstration by Wagner and colleagues in 1960 of an association between asbestos exposure and mesothelioma that lead to a resurgence of interest in mesothelial tumors and an attempt to describe their pathologic features in detail.

In the ensuing 40 years pathologists have been able to document the remarkably varied histologic appearances of diffuse malignant mesothelioma. The development of immunohistochemical stains has greatly added to the accuracy of diagnosis, and in most instances, it is possible to make a definite diagnosis of mesothelioma versus some other type of tumor. The bulk of this volume is dedicated to the description of mesothelioma, its numerous variants, and its differential diagnosis. But as the accuracy of diagnosis of malignant mesothelioma has increased, a new problem has come to the fore, namely, the clinically crucial matter of separating benign from malignant mesothelial processes. This issue, which is very complicated, has been accorded its own chapter to emphasize the importance of making this separation. At the same time, an increasing number of other primary serosal tumors and tumor-like conditions are described, and these have also been given additional space.

The ability to illustrate the numerous variants of relatively uncommon tumors depends very much on referral material from pathologists around the world, and we thank all those who have sent us such cases.

Andrew Churg, MD
Philip T. Cagle, MD
Victor L. Roggli, MD

Permission to use copyrighted illustrations has been granted by:

American Lung Association
  Am Rev Respir Dis 1975;111:12–20. For figures 1-4 and 1-5.

Elsevier
  The Developing Human: Clinically Oriented Embryology, 3rd ed, 1982. For figures 1-6 and 1-7.

McGraw-Hill
  Pathol Annu 1987;22(Pt 2):91–131. For figures 4-51 and 4-53.

Lippincott Williams & Wilkins
  Am J Surg Pathol 2002;26:1198–206. For figure 1-8.
  Am J Surg Pathol 2000;24:1183–200. For figure 5-8.

CONTENTS

1 ANATOMY, DEVELOPMENT, AND NORMAL FUNCTION OF THE SEROSAL MEMBRANES

GROSS ANATOMY

The serous membranes consist of the pleura, pericardium, peritoneum, and tunica vaginalis. The pleura forms a continuous layer over the thoracic structures, and is divided into the visceral and parietal pleurae. The visceral pleura covers the lungs and the interlobar fissures. The parietal pleura lines the thoracic wall, including the thoracic inlet, the lateral aspect of the mediastinum, the thoracic surface of the diaphragm, and forms the suprapleural membrane.

The peritoneum forms a continuous layer over the abdominal structures, except for the ostia of the oviducts. About 10 percent of individuals have a defect in the peritoneum overlying the liver. The visceral peritoneum covers the intra-abdominal organs and their mesenteries. The parietal peritoneum lines the abdominal wall, the pelvis, the undersurface of the diaphragm, and part of the anterior surfaces of retroperitoneal viscera (5). The peritoneal cavity is further divided into the supracolic, right and left infracolic, and pelvic regions. The obliquity of the mesentery of the small intestine causes the right infracolic space to taper inferiorly and the left to taper superiorly. These partitions influence the distribution of fluid within the abdominal cavity (18).

The anatomic features of the pericardium and tunica vaginalis testis conform to the same anatomic principles as those of the pleura and peritoneum. The lining of the tunica vaginalis testis is continuous with that of the peritoneal cavity.

Blood Supply

The parietal pleura derives its blood supply from the systemic circulation, i.e., the intercostal, internal mammary, and phrenic arteries. The visceral pleura is supplied by the bronchial arteries. Venous drainage runs parallel to the arterial supply. The peritoneum and mesenteries are supplied by the splanchnic arterial vessels, although a limited portion is supplied by the lower intercostal and subcostal, lumbar, and iliac arteries. Venous drainage is by the splanchnic veins and the portal system.

Lymphatic Drainage

The potential space between the visceral and parietal pleurae is drained by an extensive lymphatic system. The anterior parietal pleura drains into the intercostal lymphatic network, while the lower parietal pleura drains into the retroperitoneal nodes in the region of the kidneys and adrenal glands. The diaphragmatic pleura drains into the lower mediastinal lymph nodes. Visceral pleural lymphatic vessels drain into the pulmonary hilar lymph nodes. There is also an outward flow of lymph from the peripheral one third of the lung parenchyma towards the visceral pleura, which explains the spread of intrapulmonary tumor to the visceral pleura (15).

The peritoneal lymphatic vessels follow the corresponding blood vessels (see above). Intestinal efferent lymphatics communicate with the intramuscular and subserosal plexuses. These channels unite in the mesentery to form collecting trunks, and the lymph passes through groups of lymph nodes and "milk spots" (lymphoid aggregates in the omentum). The efferent channels ultimately form an intestinal trunk that drains the region supplied by the superior mesenteric artery and empties into the cisterna chyli or the left lumbar trunk (11). Lymph from the left side of the colon, kidneys, adrenal glands, gonads, and lower extremities is drained by the left and right lumbar lymphatic trunks (12). There are extensive communications between lymphatic vessels in the serosa on both sides of the diaphragm. Those on the right are larger and transport more fluid than those on the left.

Innervation

The parietal pleura is innervated by sensory branches from the intercostal nerves. The visceral pleura is devoid of pain fibers, but receives innervation from the vagus and sympathetic trunks. The phrenic nerve supplies the central portion of the diaphragm. The lower intercostal and abdominal wall nerves supply the peripheral portion of the diaphragm (10).

The parietal peritoneum is innervated by the spinal nerves supplying the abdominal wall. These nerves include sensory pain fibers. The visceral peritoneum receives sympathetic nerve fibers only and is devoid of pain fibers.

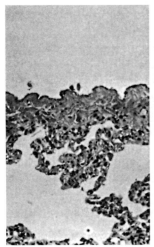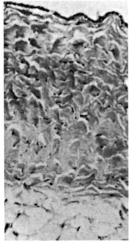

Figure 1-1

NORMAL SEROUS MEMBRANE

The submesothelial fibrous layer is much thicker in the parietal pericardium (right) than in the visceral pleura (left) (hematoxylin and eosin [H&E] stain). (Fig. 1-3 from Fascicle 20, 2nd series.)

MICROSCOPIC ANATOMY

The serous membranes are lined by a single layer of flattened mesothelial cells that rest on a basal lamina. A submesothelial layer of connective tissue of variable thickness in turn supports these cells (fig. 1-1). Mesothelial cells have abundant cytoplasm, centrally placed round nuclei, and a single small nucleolus. They range in diameter from 16 to 40 μm. The mesothelium lining the visceral pleura is characterized by more abundant microvilli than that of the parietal pleura. Mesothelial cells of the parietal pleura are separated by gaps that communicate with the underlying lymphatic vessels (20,21).

Ultrastructurally, mesothelial cells have fairly dense cytoplasm that contains ribosomes, rough endoplasmic reticulum, perinuclear tonofilaments, and moderate numbers of mitochondria. Micropinocytotic vesicles may be observed in the cell membrane. Adjacent cells are connected by relatively abundant junctional complexes and desmosomes; however, these junctions are discontinuous to permit ready diffusion of molecules between cells (9). The most prominent and characteristic ultrastructural feature of mesothelial cells is the presence of long, slender surface microvilli (figs. 1-2, 1-3). These microvilli measure up to 3.0 μm in length and 0.1 μm in diameter. They are more abundant in the caudal portions of the pleura (7), and are most abundant on the surface of organs that actively move about (2,21). It is hypothesized that the microvilli serve to entrap hyaluronate, which reduces friction at these sites.

Figure 1-2

MESOTHELIAL CELL

Branching microvilli project from the surface of the flat mesothelial cells of human pleura. (Fig. 1-4 from Fascicle 20, 2nd series.)

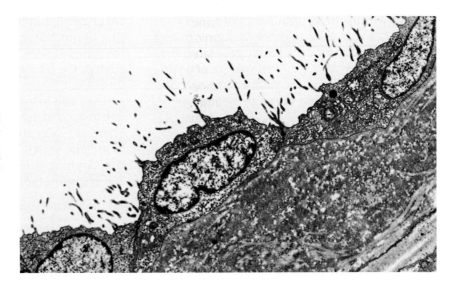

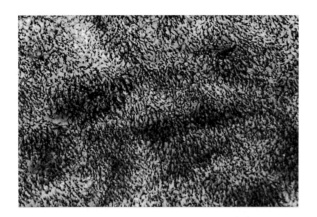

Figure 1-3

SURFACE OF MESOTHELIAL CELLS

In this scanning electron microscopic view of visceral rat pleura, the profusion of microvilli gives the surface a rough appearance. (Fig. 1-5 from Fascicle 20, 2nd series.)

Underneath the mesothelial cells lies a thin basal lamina, separating the cells from a connective tissue layer that consists of variable amounts of collagen and elastic fibers, fibroblast-like cells, capillaries, and lymphatics. Where the mesothelial cells come into contact with terminal lymphatic vessels, the basal lamina is attenuated or missing and there are gaps between cells. In the parietal pleura, these communications take the form of stomata, which allow direct continuity between mesothelium and lymphatic endothelium (figs. 1-4, 1-5) (21).

DEVELOPMENT

The intraembryonic coelomic cavity is the forerunner of the body cavities that eventually become lined by the pleural, peritoneal, and pericardial mesothelia. The intraembryonic coelomic cavity originates within the lateral and cardiogenic mesoderms as a number of isolated spaces that coalesce to form a horseshoe-shaped cavity. The curve of this cavity is the future pericardial cavity and the lateral parts are the future pleural and peritoneal cavities (17). The intraembryonic coelom divides the lateral mesoderm into the somatic and splanchnic layers that form the linings of the primitive coelomic cavity.

As development continues, the intraembryonic portion of the coelom closes off ventrally from the extraembryonic part. The area around the umbilicus is the last to close (fig. 1-6). The original right and left coelomic cavities advance

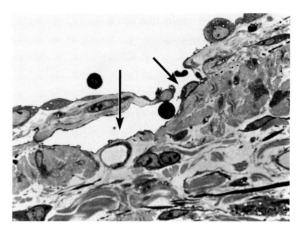

Figure 1-4

STOMA ON MESOTHELIAL SURFACE

A lymphatic lacuna (long arrow) opens onto the mesothelial surface via a narrow stoma (short arrow). The diameters of two round mononuclear cells are larger than the narrowest portion of the stoma (1 μm–thick epon section). (Fig. 9 from Wang NS. The preformed stomas connecting the pleural cavity and the lymphatics in the parietal pleura. Am Rev Respir Dis 1975;111:17.)

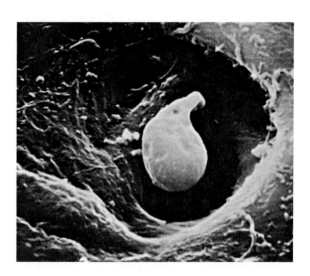

Figure 1-5

STOMA ON MESOTHELIAL SURFACE

There is a deformed red blood cell in the stoma on the subcostal parietal pleura of a rabbit. (Fig. 1 from Wang NS. The preformed stomas connecting the pleural cavity and the lymphatics in the parietal pleura. Am Rev Respir Dis 1975;111:13.)

towards each other, meeting in the midline. The lateral halves of the pericardial cavity are the first part of the intraembryonic coelom to form by the splitting of the mesoderm and the first

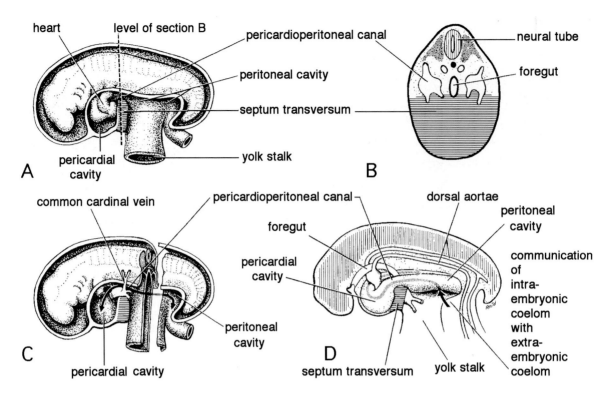

Figure 1-6

BODY CAVITIES AND MESENTERIES AT 24 DAYS' GESTATION

The drawing shows the development and relationship of the pericardial cavity, pericardioperitoneal canals, septum transversum, and intraembryonic and extraembryonic coeloms.

A: The lateral wall of the pericardial cavity has been removed to show the heart.

B: Transverse section illustrates the relationship of the pericardioperitoneal canals to the septum transversum (partial diaphragm) and the foregut.

C: Lateral view with the heart removed. The embryo has been sectioned transversely to show the continuity of the intraembryonic and extraembryonic coeloms or body cavities.

D: The pericardioperitoneal canals arise from the dorsal wall of the pericardial cavity and pass on each side of the foregut to join the peritoneal cavity. The arrow shows the communication of the extraembryonic coelom with the intraembryonic coelom and the continuity of the intraembryonic coelom. (Fig. 9-4 from Moore KL. The developing human. Clinically oriented embryology, 3rd ed. Philadelphia: WB Saunders; 1982:170.)

part completed by the union of the cavities. Layers of splanchnic mesoderm couple in the mid-portion of the embryo to form the various mesenteries. Mesoderm ventral to the gut attenuates, and the cavities coalesce.

The septum transversum partitions the single coelomic space into thoracic and abdominal portions, and is joined laterally by the pleuroperitoneal folds that arise from the body wall. Bilateral pleuropericardial folds then partition the pleural from the pericardial cavities and eventually join the septum transversum (3). The septum transversum arises in the cervical area of the embryo, moving caudad to the level of the first lumbar vertebra by the second month of gesta-

tion. As it descends, it carries nerve fibers from the phrenic nerves. The lungs also begin their development in a cephalad location, arising as buds from the endoderm that projects into the early pleuropericardial folds from the mesodermal ridge on each side (fig. 1-7) (3).

The pleural and pericardial cavities complete their separation first on the right side. On the left side, a communication between the pleural and pericardial cavities usually persists (14). The diaphragm derives from the septum transversum (which forms the central tendon), the two pleuroperitoneal membranes, the mesentery of the esophagus, and the dorsal and lateral body walls (which form the muscular portion).

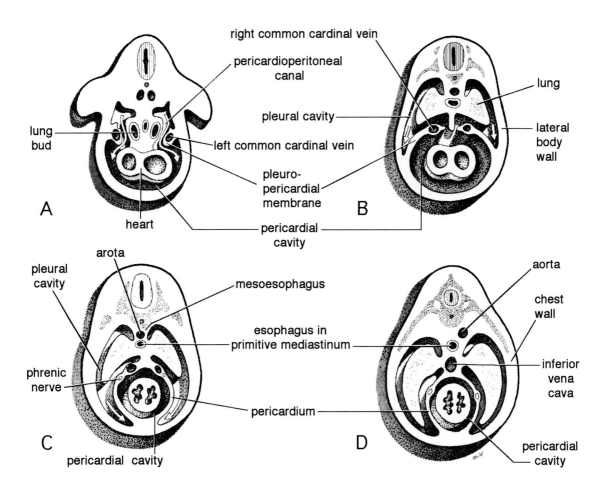

Figure 1-7

STAGES IN THE DEVELOPMENT OF THE PLEURAL CAVITIES

Schematic drawing of a transverse section through an embryo cranial to the septum transversum illustrates successive stages in the separation of the pleural cavities from the pericardial cavity. Growth and development of the lungs, expansion of the pleural cavities, and formation of the fibrous pericardium are also shown.

A: Five weeks. The arrows indicate the communication between the pericardioperitoneal canals and the pericardial cavity.

B: Six weeks. The arrows indicate development of the pleural cavities as the pericardioperitoneal canals extend and the pleural cavities expand into the body wall.

C: Seven weeks. The pleural cavities expand ventrally around the heart. The pleuropericardial membranes are now fused in the midline with each other and with the mesoderm ventral to the esophagus.

D: Eight weeks. Continued expansion of the lungs and pleural cavities and formation of the fibrous pericardium and chest wall are illustrated. (Fig. 9-5 from Moore KL. The developing human. Clinically oriented embryology, 3rd ed. Philadelphia: WB Saunders; 1982:171.)

The parietal pleura, pericardium, and peritoneum derive from somatic mesoderm, whereas the visceral membranes derive from splanchnic mesoderm. These membranes consist of a layer of persisting mesenchymal tissue and a layer of flattened surface cells that become the mesothelium. These layers retain the potential to differentiate in a number of ways (14).

NORMAL FUNCTION

The purpose of the serosal lining is to provide an essentially frictionless surface over which the viscera can move as they perform their contractile and expansile functions. A thin layer of liquid that is rich in hyaluronate coats the serosal surfaces, and these surfaces are actively involved in fluid transport (19). The

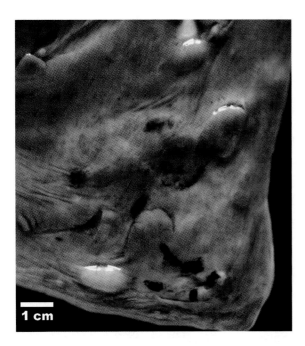

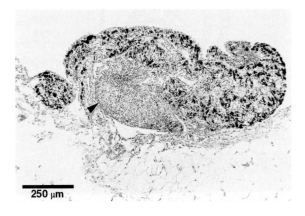

Figure 1-8

BLACK SPOTS ON THE PARIETAL PLEURA

Top: Pigmented spots observed at autopsy in the parietal pleura range from 3 to 10 mm in diameter.

Bottom: Hematoxylin and eosin–stained section of a pigmented spot. The arrow indicates the area with a chronic inflammatory reaction. (Figs. 1D and 2A from Mitchev K, Dumortier P, De Vuyst P. 'Black spots' and parietal pleural plaques on the parietal pleura of 150 urban necropsy cases. Am J Surg Pathol 2002;26:1200–1201.)

visceral and pleural surfaces are separated by a distance of 5 to 10 μm. This space is occupied by the glycosaminoglycan lubricant, hyaluronate, and cells, including pleural macrophages (8).

Fluid moves from the blood stream to the pleural space and vice versa by means of hydrostatic and colloidal osmotic pressures in the serosal capillaries. Transudation of fluid from the arterial ends of the capillaries in the parietal pleura and reabsorption in the venous ends of capillaries in the visceral pleura is favored, since the hydrostatic pressure in the systemic circulation supplying the former is greater than the pressure in the pulmonary circulation supplying the latter (6).

Approximately 10 to 20 percent of pleural fluid is reabsorbed by the lymphatic vessels of the parietal pleura. Larger molecules (e.g., proteins) enter directly into the lymphatic channels. Particulate matter and cells are also taken up by the parietal pleura, principally through the stomata of the lower mediastinal and costal areas (fig. 1-8) (1,6,16,19), but not by the visceral pleura. Particulate matter in the peritoneal cavity is reabsorbed through diaphragmatic lymphatic vessels. The flow of lymph is influenced by muscular activity during respiration (4).

Obstruction of lymph flow may result in the accumulation of fluid in the pleural, pericardial, or peritoneal spaces. Lymph flow can be effectively blocked by tumor, fibrosis, or particulate matter. The proteinaceous content of accumulated fluid can augment the retention of additional fluid by means of its oncotic pressure. Pleural thickening (e.g., diffuse pleural fibrosis) may seal lymphatic vessels and thus prevent the absorption of fluid. Overload of the lymphatic drainage system in the peritoneal cavity may result in retention of fluid in the thorax (particularly on the right), since ascitic fluid drains through the diaphragm into the pleural cavity (4).

The relative importance of the various compartments within the mesothelial cell for fluid transport is incompletely understood and somewhat controversial. Fluid and particle transmission may occur through pinocytotic vesicles or directly through the cytosol. Long surface microvilli increase the surface area and thus may promote fluid transport. Pinocytosis may be more important for equalization of protein and salt composition of the serous fluid (13). Some have suggested that intercellular clefts are the major sites of solute and fluid transport across the peritoneal membrane (9).

REFERENCES

1. Albertine KH, Wiener-Kronish JP, Staub NC. The structure of the parietal pleura and its relationship to pleural liquid dynamics in sheep. Anat Rec 1984;208:401–9.
2. Andrews PM, Porter KR. The ultrastructural morphology and possible functional significance of mesothelial microvilli. Anat Rec 1973;177:409–26.
3. Arey LB. Developmental anatomy; a textbook and laboratory manual of embryology. Rev. 7th ed. Philadelphia: Saunders; 1974:284–94.
4. Battifora H, McCaughey WT. Tumors of the serosal membranes. Atlas of Tumor Pathology, 3rd Series, Fascicle 15. Washington, DC: Armed Forces Institute of Pathology; 1995:1–7.
5. Bender MD, Ockner RK. Diseases of the peritoneum, mesentery, and diaphragm. In: Sleisenger MH, Fordtran JS, eds. Gastrointestinal disease: pathology, diagnosis, management, 2nd ed. Philadelphia: Saunders; 1978:1947–8.
6. Black LF. The pleural space and pleural fluid. Mayo Clin Proc 1972;47:493–506.
7. Carter D, True L, Otis C. Serous membranes. In: Sternberg SS, ed. Histology for pathologists. New York: Raven Press; 1992:499–514.
8. Choe N, Tanaka S, Xia W, Hemenway DR, Roggli VL, Kagan E. Pleural macrophage recruitment and activation in asbestos-induced pleural injury. Environ Health Persp 1997;105(Suppl 5):1257–60.
9. Cotran RS, Karnovsky MJ. Ultrastructural studies on the permeability of the mesothelium to horseradish peroxidase. J Cell Biol 1968;37:123–37.
10. Green RA, Johnston RF. Introduction to pleural disease. In: Baum GL, ed. Textbook of pulmonary diseases, 2nd ed. Boston: Little, Brown; 1974:941–57.
11. Haagensen CD. General anatomy of the lymphatic system. In: Haagensen CD, ed. The lymphatics in cancer. Philadelphia: Saunders; 1972:22–41.
12. Hollingshead WH. Textbook of anatomy, 3rd ed. Hagerstown, MD: Harper & Row; 1974:640.
13. Ivanova VF. [Role of the mesothelium of the parietal peritoneum in the process of absorption of true solutions and India ink suspension from the abdominal cavity.] Biull Eksp Biol Med 1976;82:1258–61. (Russian)
14. Langman J. Coelomic cavity and mesenteries. In: Langman J. Medical embryology, 3rd ed. Baltimore: Williams & Wilkins; 1975:303–17.
15. Lowell JR. Anatomy and physiology. In: Pleural effusions: a comprehensive review. Baltimore: University Park Press; 1977:7–11.
16. Mitchev K, Dumortier P, De Vuyst P. 'Black spots' and hyaline pleural plaques on the parietal pleura of 150 urban necropsy cases. Am J Surg Pathol 2002;26:1198–206.
17. Moore KL. Coelomic cavity and mesenteries. In: Moore KL, ed. The developing human: clinically oriented embryology. Philadelphia: Saunders; 1982:303–17.
18. Morson BC. The peritoneum. In: Symmers WS, ed. Systemic pathology, 2nd ed. London: Churchill Livingstone; 1978:1179–97.
19. Pistolesi M, Miniati M, Giuntini C. Pleural liquid and solute exchange. Am Rev Respir Dis 1989;140:825–47.
20. Wang NS. The preformed stomas connecting the pleural cavity and the lymphatics in the parietal pleura. Am Rev Respir Dis 1975;111:12–20.
21. Wang NS. The regional difference of pleural mesothelial cells in rabbits. Am Rev Respir Dis 1974;110:623–33.

2 CLASSIFICATION OF TUMORS OF THE SEROSAL MEMBRANES

There is no ideal classification system for tumors of the serosal membranes. The histogenesis of many of these lesions is obscure, with arguments in the literature about an origin from mesothelial cells, submesothelial mesenchymal cells, and uncommitted stem cells. Mesothelial cells display a remarkable morphologic plasticity, manifested by their ability to develop as a wide variety of epithelial and spindle forms that in some instances perfectly recapitulate named tumors elsewhere in the body (e.g., the occurrence of osteosarcomatous foci in malignant mesothelioma). This plasticity further complicates the problem, as does the occurrence in the serosal membranes of tumors that probably have no specific relationship to serosal cells since they are found elsewhere in the body. Solitary fibrous tumors, diffuse angiosarcomas, and synovial sarcomas of the serosal membranes epitomize this problem. Thus, the relationship, if any, of many serosal tumors to each other remains obscure.

For that reason we have classified serosal tumors first into diffuse and localized forms, and secondarily into localized tumors that are benign (or, for some well-differentiated papillary mesotheliomas, of uncertain malignant potential) (Table 2-1). The rationale behind this approach is simply that it provides excellent prognostic information. To our knowledge, all diffuse neoplasms of the serosal membranes, with the exception of diffuse forms of well-differentiated papillary mesothelioma, are malignant, and most, with the exception of some cases of peritoneal mesothelioma, are at present incurable. On the other hand, patients with even malignant forms of localized serosal tumors may have a very good long-term prognosis if the tumor is completely excised, even though the tumor may be microscopically identical to a diffuse tumor. Localized malignant mesothelioma and localized synovial sarcomas are excellent examples of this phenomenon.

Table 2-1

CLASSIFICATION OF TUMORS OF THE SEROSAL MEMBRANES

Diffuse Tumors of the Serosal Membranes
Diffuse malignant mesothelioma
Carcinoma, sarcoma, and other tumors metastatic to the pleura and peritoneum
Serous papillary carcinoma and serous papillary borderline tumor of the peritoneum
True primary sarcomas of the pleura and peritoneum
  Vascular sarcomas
  Synovial sarcomas
  Other defined sarcomas
Squamous carcinomas

Localized Benign Tumors, Tumors of Uncertain Malignant Potential, and Tumor-Like Conditions of the Serosal Membranes
Solitary fibrous tumor
Nodular pleural plaque
Well-differentiated papillary mesothelioma
Lipoma
Adenomatoid tumor
Calcifying fibrous tumor
Simple mesothelial cysts of the peritoneum
Multicystic mesothelioma of the peritoneum
Schwannoma

Localized Malignant Tumors of the Serosal Membranes
Malignant version of solitary fibrous tumor
Localized malignant mesothelioma of pleura or peritoneum
Desmoplastic small round cell tumor
Askin's tumor
Pleuropulmonary blastoma
Liposarcoma
Lymphoma

Miscellaneous Conditions
Endometriosis and endosalpingiosis
  Leiomyomatosis peritonealis disseminata
  Gliomatosis peritonei
Deciduosis
Chondroid and osseous metaplasia
Trophoblastic implants
Omental-mesenteric myxoid hamartoma

An important corollary of this approach to classification is the necessity of obtaining clinical information before making a diagnosis. Whether a given tumor is diffuse or localized is often not apparent from the pathology specimen. This information needs to be obtained elsewhere, sometimes from the surgical pathology requisition, but often from radiographic reports, or from descriptions of the findings at thoracoscopy/thoracotomy or laparoscopy/laparotomy. This information may suggest that the tumor in question really is primary in the serosal membranes, rather than representing metastasis or direct spread from another site.

3 CYTOLOGY OF THE SEROSAL SURFACES

UTILITY AND LIMITATIONS OF CYTOLOGY OF THE SEROSAL SURFACES—CYTOLOGY VERSUS HISTOLOGY: GENERAL OVERVIEW

Exfoliative Cytology of Body Cavity Fluids

Exfoliative cytologic examination of effusions of the pleural cavity, peritoneal cavity (ascites), and pericardial cavity is a common method for determining whether an effusion is benign or malignant in origin. The long list of possible etiologies of effusions means that many fluids examined by the cytopathologist will be benign, and must be differentiated from malignancy, a task that can be difficult in some situations.

Effusions are subdivided into transudates and exudates based on the ratio of effusion protein to serum protein, an evaluation determined in the clinical laboratory. Transudates result from alterations in hydrostatic or oncotic pressure, often due to systemic factors, that cause physiologic imbalances in the formation and reabsorption of fluid. The most common cause of transudative pleural effusion is congestive heart failure. Exudates result from pathologic processes localized to the serosal membranes and adjacent tissues that produce leakage of proteins and cells from damaged capillaries. Exudates have a higher protein content and cellularity than transudates. The most common cause of exudative pleural effusions in the United States is infection, followed by pleural malignancy (approximately 200,000 pleural metastases and 1,500 diffuse malignant mesotheliomas each year) (29,32).

Cells exfoliated into effusion fluid can be examined as cytology smears, liquid-based preparations, cytospin preparations, or cell blocks under the light microscope. Effusions containing exfoliated cancer cells are exudates. Malignancies, however, can indirectly cause benign transudates or benign exudates by blocking pleural lymphatic vessels or by causing peritumoral reactions in overlying pleural tissues without shedding cancer cells into the fluid. Therefore, an exfoliative cytology specimen that is negative for cancer cells does not necessarily rule out a malignancy. Only 30 to 40 percent of pleural diffuse malignant mesotheliomas and 60 to 70 percent of cancers metastatic to the pleura are diagnosed by exfoliative cytology (64). Diagnostic dilemmas occur when exfoliative cytology specimens contain benign reactive mesothelial cells that mimic malignant cells or contain abundant blood or inflammatory cells that obscure the malignant cells in the fluid (3, 14,29,32,34,40,43,49,52).

Needle Aspiration Cytology

Cytology specimens and small cores of tissue can be obtained by needle aspiration, but sampling error and small specimen size often limit the ability to make a diagnosis. Whereas thoracoscopic biopsy is positive in more than 90 percent of patients with pleural cancer, needle aspiration biopsy is positive in less than 25 percent of patients with cancer who have a negative effusion cytology (3,49).

Exfoliative Cytology and Aspiration Cytology Versus Histology

Exfoliative cytology and aspiration cytology are of limited usefulness in diagnosing diffuse malignant mesothelioma. As discussed further below, benign reactive mesothelial cells may have cytologic features that mimic malignancy and diffuse malignant mesotheliomas may be cytologically bland. Therefore, without evidence of invasion of underlying tissues provided by an adequate histologic sample, a diagnosis of mesothelioma on purely cytologic grounds is difficult or impossible in many situations. Sarcomatous mesotheliomas typically do not shed cells into effusion fluid. A small biopsy that does not allow assessment of underlying invasion may also be of limited usefulness in diagnosing malignancy (see chapters 4 and 5) (3,49).

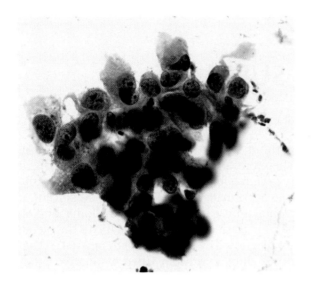

Figure 3-1

CYTOLOGIC FEATURES OF MALIGNANCY

These nonsmall cell carcinoma cells are arranged in a cohesive nest.

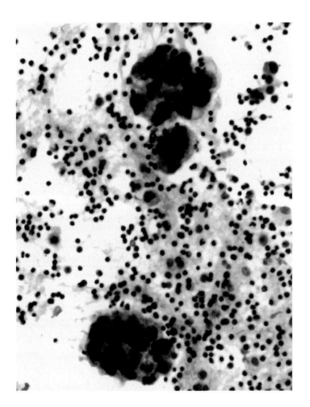

Figure 3-2

CYTOLOGIC FEATURES OF MALIGNANCY

The breast carcinoma cells are arranged in ball-like, three-dimensional, cohesive clusters called morulae. (Courtesy of Dr. T. Allen, Tyler, TX.)

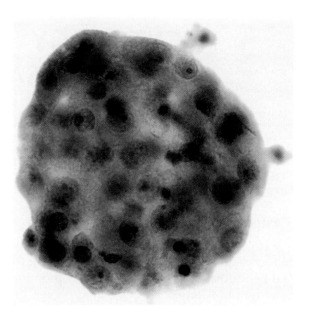

Figure 3-3

CYTOLOGIC FEATURES OF MALIGNANCY

High-power microscopy shows the three-dimensional character of a morula in diffuse malignant mesothelioma.

BENIGN VERSUS MALIGNANT MESOTHELIAL PROLIFERATIONS IN EFFUSION CYTOLOGY

Cytologic Features of Malignancy

Cancer cells are present in malignant effusions as individual cells, sheets of cohesive cells, and three-dimensional spherical clusters of cohesive cells called morulae (figs. 3-1–3-3). The three-dimensional structure and compact cellularity of the morulae can be observed by focusing the microscope up and down on the cluster of cells. Cancer cells, particularly adenocarcinoma cells, may be arranged in papillary or acinar structures similar to their architecture on histologic samples (figs. 3-4–3-6). Although hypercellularity or increased numbers of cells representing the cancer cells is considered a feature of malignancy in exfoliative cytology specimens, cancer cells may be few in number. Psammoma bodies (fig. 3-7) can be seen with papillary cancers including papillary adenocarcinoma and papillary diffuse malignant mesothelioma (3,13,14,34,40,49,52).

The classic cytologic features of individual cancer cells in malignant effusions are enlarged cells

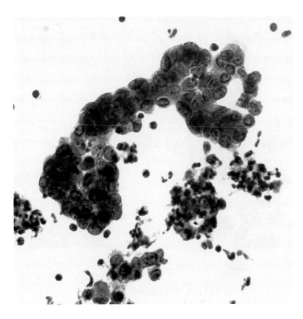

Figure 3-4

CYTOLOGIC FEATURES OF MALIGNANCY

Cells of a serous papillary carcinoma of the ovary are arranged in a three-dimensional papillary frond.

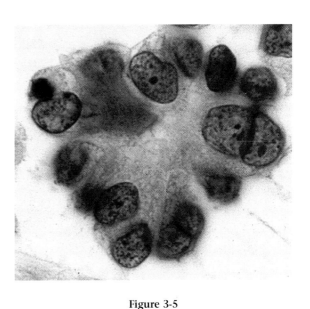

Figure 3-5

CYTOLOGIC FEATURES OF MALIGNANCY

These adenocarcinoma cells are arranged in an acinar structure as if around a lumen, reminiscent of a gland in a histology specimen. (Courtesy of Dr. M. L. Ostrowski, Houston, TX.)

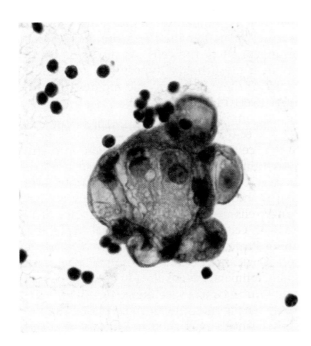

Figure 3-6

CYTOLOGIC FEATURES OF MALIGNANCY

The cells of this mucinous adenocarcinoma of the gastrointestinal tract have cytoplasmic mucin and are in an acinar-like arrangement.

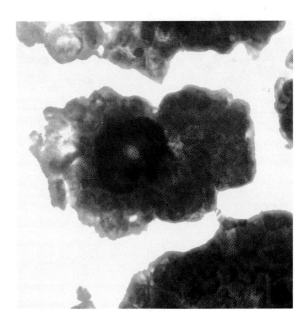

Figure 3-7

CYTOLOGIC FEATURES OF MALIGNANCY

A psammoma body is seen in the center of a papilla of serous papillary carcinoma of the ovary.

13

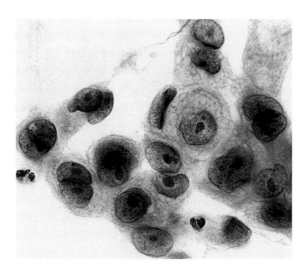

Figure 3-8

CYTOLOGIC FEATURES OF MALIGNANCY

The adenocarcinoma cells have enlarged nuclei with irregular borders and large irregular nucleoli. (Courtesy of Dr. M. L. Ostrowski, Houston, TX.)

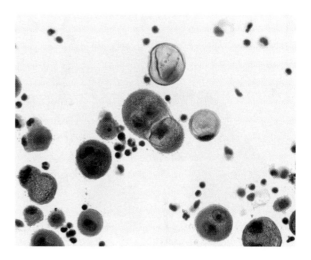

Figure 3-10

REACTIVE MESOTHELIAL CELLS

Individual reactive mesothelial cells have an oval nucleus and abundant cytoplasm bordered by a fuzzy rim. They are surrounded by more numerous lymphocytes and histiocytes. A characteristic "window" or space is seen between mesothelial cells.

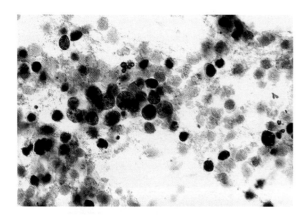

Figure 3-9

CYTOLOGIC FEATURES OF MALIGNANCY

This small cell carcinoma shows cytologically characteristic small cells and numerous necrotic tumor cells.

with enlarged nuclei (high nucleus to cytoplasm ratio), coarse chromatin, enlarged and multiple nucleoli, and irregular or indented nuclear contours (fig. 3-8). Mitoses, atypical mitoses, and necrotic debris are also typical features of cancers in cytology specimens (fig. 3-9). These cytologic features are generally characteristic of cancer cells, but specific cytologic features may reflect the cell type of the cancer: for example, adenocarcinomas may display intracytoplasmic

mucin vacuoles and melanomas may contain melanin pigment. The diagnosis of cancer is based on a combination of individual cytologic features of malignancy and the architectural arrangement of the cells. As discussed below, atypical mesothelial hyperplasia can mimic many of these cytologic and architectural features (3,13,14,34,40,49,52).

Reactive Atypia of Mesothelial Cells

Benign mesothelial cells exfoliate into effusion fluid in many situations and display a spectrum of reactive change that range from minimal simple reactive change to highly atypical reactive change, mimicking malignancy (3,14, 26,34,49). Both simple mesothelial hyperplasia and reactive atypical mesothelial hyperplasia result in increased cellularity of exfoliative specimens, a general feature already noted to be typical of many cancer specimens (figs. 3-10, 3-11). With simple mesothelial hyperplasia, the mesothelial cells may shed in clusters or sheets, with adjacent cells separated from one another by spaces traditionally referred to as "windows" (fig. 3-12). The exfoliated cells of simple hyperplasia maintain recognizable morphologic features of normal mesothelial cells, including round to oval nuclei with thin distinct nuclear

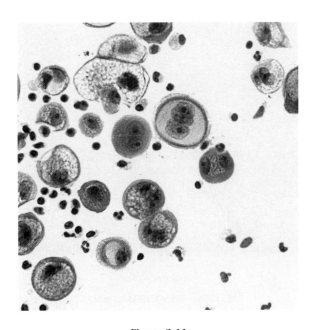

Figure 3-11

REACTIVE MESOTHELIAL CELLS

The multinucleate cell in the center shows paler inner cytoplasm with a dense ring of peripheral cytoplasm and an outer fuzzy rim.

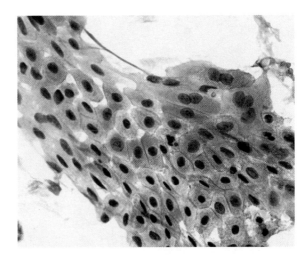

Figure 3-12

REACTIVE MESOTHELIAL CELLS

Reactive mesothelial cells often shed in sheets in peritoneal washings and fine-needle aspirates. These sheets are composed of cells with round to oval nuclei, dense cytoplasm, sharp cell borders, and occasional windows between cells.

membranes; vesicular to finely granular chromatin; round uniform nucleoli; and abundant, dense, darkly staining cytoplasm. The peripheral cytoplasm often stains darker than the central cytoplasm and microvilli around the periphery of the cell result in a characteristic fuzzy rim or border. Cytoplasmic vacuoles may be present and should be differentiated from the mucin-containing vacuoles of adenocarcinoma cells. Binucleation or multinucleation is not uncommon and nucleoli may be multiple within a single nucleus. When multinucleation is present, the different nuclei within the same benign cell appear essentially identical.

Reactive mesothelial cells with greater degrees of cytologic atypia are more difficult to differentiate from cancer cells. Atypical reactive cytologic features include enlarged nuclei, coarse chromatin, prominent nucleoli, and frequent mitoses, features usually associated with malignancy. Nuclear contours may show more variation than simple hyperplasia, but are still generally round to oval, with a smooth nuclear membrane. Cytoplasmic vacuoles may compress the nucleus, suggesting the signet ring cells of adenocarcinoma. Atypical mesothelial cells

can also architecturally mimic malignancies, exfoliating into the effusion fluid as papillary excrescences or forming rosettes or three-dimensional clusters within the fluid. Acinar structures composed of reactive mesothelial cells have been reported in 6 percent of benign effusions (figs. 3-13, 3-14). The combination of increased cellularity, individual cell cytologic atypia, and similar architectural features overlaps with the features usually associated with cancer (3,13,14, 34,40,49,52).

False Positives and False Negatives in Serosal Cavity Cytology

There are many more nonmalignant effusions than malignant effusions, since only approximately 15 percent of pleural effusions in the United States each year are caused by cancer. Of these malignant pleural effusions, less than 1 percent are caused by diffuse malignant mesothelioma (29,32). Due to the overlap of cytologic features of benign and malignant cells, it is often not possible to reliably determine whether mesothelial cells are benign or malignant in a cytology specimen (fig. 3-15). Merely determining whether or not cells are mesothelial does not distinguish between benign and

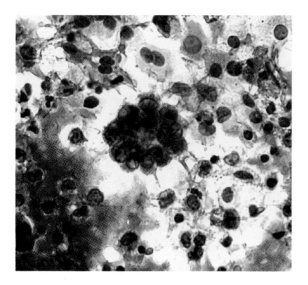

Figure 3-13

ATYPICAL MESOTHELIAL CELLS

This cluster of mesothelial cells resembles an acinar structure in a specimen that otherwise had features compatible with a reactive process. Reactive mesothelial cells may shed in acinar or gland-like structures.

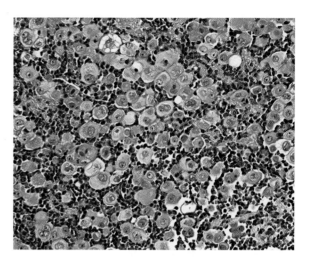

Figure 3-15

ATYPICAL MESOTHELIAL CELLS

The florid cellularity consists of mesothelial cells with cytologic atypia in a hemorrhagic background. This may be seen as reactive hyperplasia, particularly in the setting of pulmonary infarct, which may produce a bloody background. Diffuse malignant mesothelioma composed of well-differentiated tumor cells cannot be excluded (hematoxylin and eosin [H&E]-stained cell block).

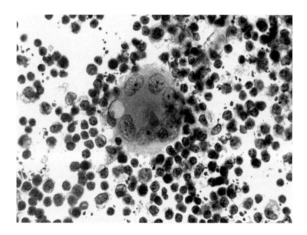

Figure 3-14

ATYPICAL MESOTHELIAL CELLS

The mesothelial cells have cytologic features compatible with reactive cytologic atypia and are arranged in an acinar-like structure.

malignant mesothelial cells. Therefore, it is necessary to first determine if cells in a fluid are truly malignant before secondarily determining the type of cancer they represent (fig. 3-16). In addition to reactive mesothelial cells, cells of nodular histiocytic hyperplasia should not be confused with malignant cells (11).

The inability to assess for invasion is one of the potential drawbacks of exfoliative cytology in the diagnosis of diffuse malignant mesothelioma. Reactive mesothelial cells stain for the same markers of mesothelial differentiation as do diffuse malignant mesothelioma cells (cytokeratin [CK]7, CK5/6, calretinin, Wilms' tumor 1 [WT1], mesothelin). Thus, confirming that atypical cells are mesothelial in origin does not determine whether the cells are benign or malignant (8,14,16,22,27,33,37,39,41–43,54,59, 64,67). Care should also be taken in interpreting immunostains of cytology specimens, since immunostaining cytospin preparations is technically more difficult than immunostaining cell blocks and histologic sections, and can produce false staining patterns (16,49,57). For the same reason, a diagnosis of malignancy in exfoliated cells should not be based purely on apparently positive immunostains for carcinoma markers.

The presence of regular round to oval nuclei with smooth nuclear membranes favors a benign mesothelial cell. Generally, atypical reactive mesothelial cells blend with cells with lesser degrees of reactive atypia within a benign effusion, giving the impression of one population

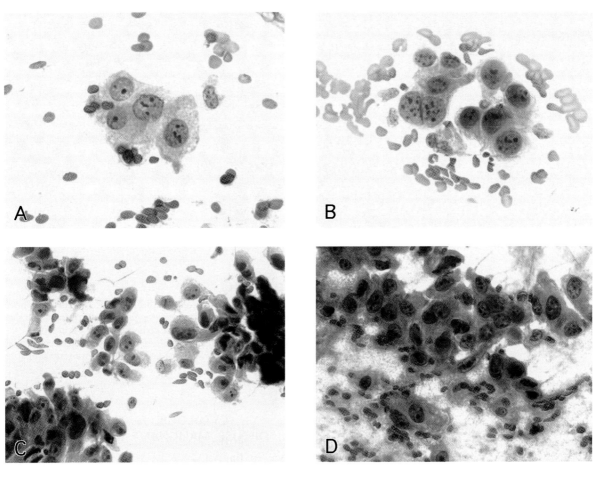

Figure 3-16

ATYPICAL MESOTHELIAL CELLS

A: The cells are enlarged, have round to oval nuclei with regular borders, and have conspicuous nucleoli. These features are most consistent with reactive cytologic atypia, although a diffuse malignant mesothelioma composed of well-differentiated tumor cells cannot be excluded (H&E-stained cell block.) (Figures A through E are from the same case.)

B: Most of the cells have features that, by themselves, could be compatible with reactive cytologic atypia, but the presence of a binucleate cell with coarse chromatin is worrisome for malignancy. On the basis of these cells alone, a definite diagnosis cannot be rendered.

C: The well-preserved cells in the center have cytologic features compatible with reactive cytologic atypia. The cells at the periphery have dark, irregular nuclei, a feature that could be the result of crush and other artifacts. On the basis of these cells alone, a definite diagnosis cannot be rendered.

D: These cells show features of malignancy: irregular nuclear shape and irregular chromatin, prominent nuclei, and enlarged cells. Severe reactive atypia, however, cannot be excluded on the basis of these cells alone.

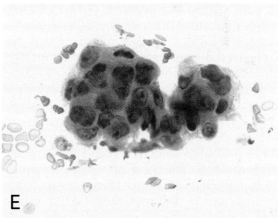

E: These cells are the most typical of malignancy. The cells form a ball-like cluster or morula, have irregular nuclei and irregular chromatin, prominent nuclei, and a mitosis. The presence of a mitosis alone does not diagnose malignancy since mitoses occur in benign reactive hyperplasia. The "scalloping" between cells is characteristic of clusters of mesothelial cells, whether benign or malignant, and helps differentiate mesothelial cells from clusters of adenocarcinoma cells which typically lack this scalloping.

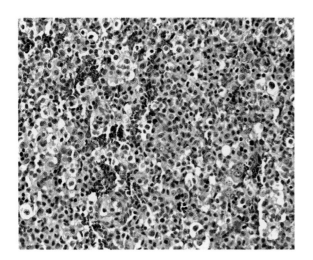
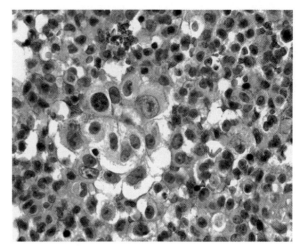

Figure 3-17

ATYPICAL MESOTHELIAL CELLS

Left: At low-power magnification, there is high cellularity of mesothelial cells mixed with inflammatory cells and blood. At this magnification, all that can be said is that there is hypercellularity. These cells could represent benign florid reactive hyperplasia, but a diffuse malignant mesothelioma composed of well-differentiated tumor cells cannot be excluded.

Right: Higher-power magnification shows a spectrum of cells, from smaller mesothelial cells, to somewhat larger mesothelial cells that by themselves could be compatible with reactive cytologic atypia, to a cluster of very large cells that are probably malignant mesothelial cells (H&E-stained cell block).

of cells with variable degrees of atypia (fig. 3-17). When a cancer is present, the reactive mesothelial cells and the frankly malignant cells often appear as two separate and distinct populations of cells within the effusion (3,14,26,34,49). Subtle qualitative differences in otherwise overlapping features of benign and malignant mesothelial cells may also favor one diagnosis over the other, but caution is advised (fig. 3-18).

Other factors can confound the diagnosis of diffuse malignant mesothelioma by cytology. As discussed, some epithelial diffuse malignant mesotheliomas, the type most likely to exfoliate into effusion fluid, have bland cytologic features that resemble those of benign hyperplastic mesothelial cells. Since there are early stage mesotheliomas, and mesothelial cell dysplasia and mesothelioma in situ likely exist, some atypical mesothelial proliferations in effusions may represent these early stages of mesothelioma even before a tumor mass is clinically detectable. Also, benign effusions with benign reactive mesothelial cells can result from obstruction of lymphatic vessels by cancer or reactive serosal changes overlying a cancer without the shedding of cancer cells into the fluid.

CYTOLOGIC FEATURES OF DIFFUSE MALIGNANT MESOTHELIOMA

Epithelial Diffuse Malignant Mesothelioma

Diffuse malignant mesotheliomas cause less than 1 percent of malignant pleural effusions, and only epithelial diffuse malignant mesotheliomas are likely to exfoliate cells into effusion fluids. Epithelial malignant mesothelioma cells often have a recognizably mesothelial appearance, and combine the cytologic features of benign mesothelial cells described above with the cytologic features associated with malignancy (fig. 3-19) (3–5,13,23,25,26,46,49,53,58, 60,63). Epithelial malignant mesothelioma cells typically lack the significant degree of cytologic pleomorphism seen with carcinoma cells and they can be bland. Architecturally, these cells are arranged in sheets, clusters, morulae, or papillary structures, potentially mimicking adenocarcinoma or other cancers and overlapping with reactive atypical mesothelial hyperplasia (figs. 3-20, 3-21). Similar to other papillary malignancies, papillary malignant mesotheliomas can form psammoma bodies. Epithelial malignant mesothelioma cells may have cytoplasmic

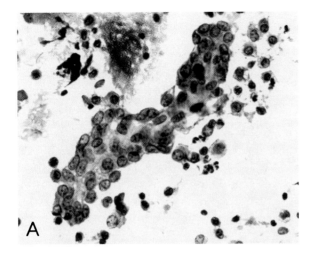

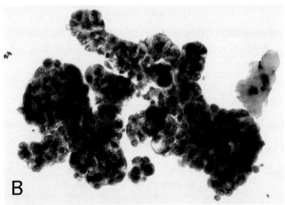

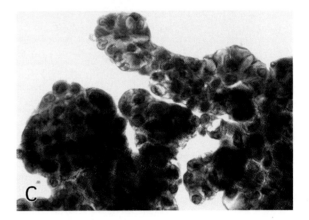

Figure 3-18

REACTIVE MESOTHELIAL CELLS

A: Reactive mesothelial cells may form papillary fronds that suggest papillary malignancy. The cells lack the frank cytologic features of malignancy, however, and the fronds tend to lack the three-dimensional character of papillae of malignancies.

B: These papillary fronds are very worrisome for papillary diffuse malignant mesothelioma because of the three-dimensional character with overlapping cells.

C: On higher-power microscopy, evaluation of the cytologic features of individual cells is difficult because the cells overlap, emphasizing the three-dimensional structure of the papillae. This is highly suspicious for diffuse malignant mesothelioma rather than a papillary frond shed from reactive hyperplasia.

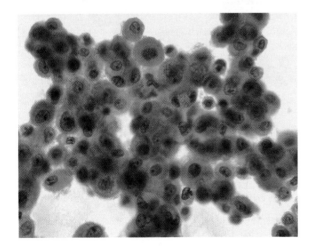

Figure 3-19

EPITHELIAL DIFFUSE MALIGNANT MESOTHELIOMA

The cells have features of mesothelial cells including oval nuclei with nucleoli, dense cytoplasm, and a darker peripheral rim with a fuzzy outer border. There are windows between the cells. The cells are slightly enlarged and increased in number. These cells are difficult to distinguish from those of reactive mesothelial hyperplasia.

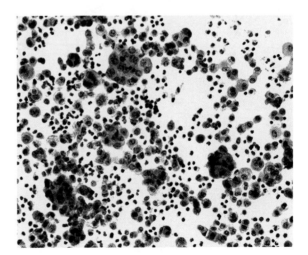

Figure 3-20

EPITHELIAL DIFFUSE MALIGNANT MESOTHELIOMA

Three-dimensional clusters, or morulae, of mesothelial cells are mixed with individual tumor cells and inflammatory cells.

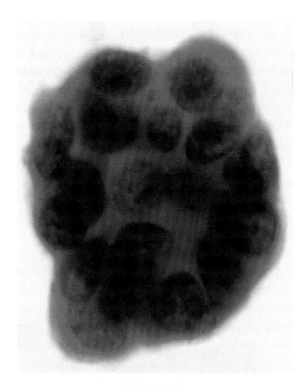

Figure 3-21

EPITHELIAL DIFFUSE MALIGNANT MESOTHELIOMA

The three-dimensional morula is composed of cytologically atypical mesothelial cells with prominent nucleoli.

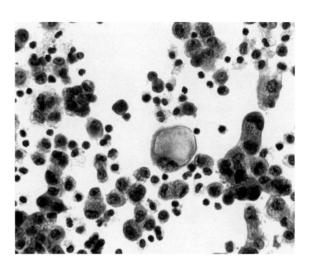

Figure 3-22

EPITHELIAL DIFFUSE MALIGNANT MESOTHELIOMA

A cytoplasmic vacuole pushes the nucleus of the tumor cell to one side.

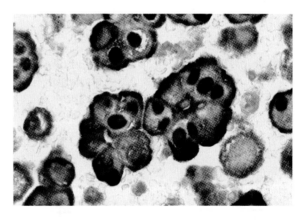

Figure 3-24

EPITHELIAL DIFFUSE MALIGNANT MESOTHELIOMA

The mesothelioma cells in a pleural fluid cell block show cytoplasmic and strong nuclear immunostaining for calretinin. (Courtesy of Dr. R. Laucirica, Houston, TX.)

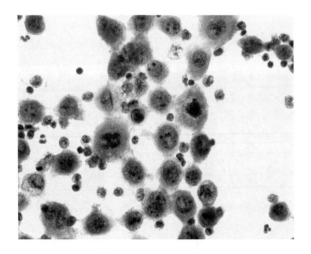

Figure 3-23

EPITHELIAL DIFFUSE MALIGNANT MESOTHELIOMA

Two mitotic figures are seen in the same high-power field. Although mitoses are a feature often seen in malignancy, their presence alone is not a basis for a diagnosis of diffuse malignant mesothelioma since benign reactive hyperplasia may also have mitoses.

vacuoles and even form "signet ring"–like cells that mimic the mucin-filled cytoplasmic vacuoles of adenocarcinoma (figs. 3-22, 3-23). In these settings, histochemical stains for mucins and immunostains of sections of cell block preparations can be used, with caution, to confirm that malignant cells are of mesothelial origin (figs. 3-24–3-26) (14,16,27,33,37,39,43,54,64).

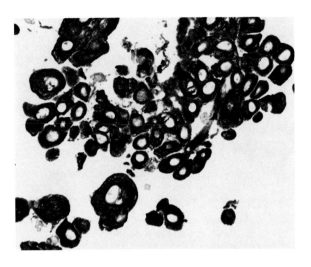

Figure 3-25

EPITHELIAL DIFFUSE MALIGNANT MESOTHELIOMA

Mesothelioma cells show strong cytoplasmic immuno-positivity for cytokeratin (CK) 5/6 in this pleural fluid cell block.

Sarcomatous Diffuse Malignant Mesothelioma

Sarcomatous mesotheliomas cause effusions by local effects on serosal membranes and obstructing lymphatic vessels. They usually do not exfoliate cancer cells into the effusion fluid and, therefore, effusions caused by sarcomatous mesothelioma are typically not malignant effusions and cannot be diagnosed by exfoliative cytology. The exclusion of most sarcomatous mesotheliomas from diagnosis is yet another limitation of exfoliative cytology in the diagnosis of diffuse malignant mesothelioma. With fine-needle aspiration cytology, sarcomatous mesotheliomas consist of malignant spindle cells for which the differential diagnosis is primarily sarcoma and sarcomatoid carcinoma (fig. 3-27) (3,29,32,49).

Mixed Epithelial and Sarcomatous Diffuse Malignant Mesothelioma

Biphasic mesotheliomas may shed their epithelial component but are unlikely to shed their sarcomatous component into effusion fluid. The exfoliated epithelial component has cytologic characteristics associated with purely epithelial mesothelioma (3,49,53). Rarely, transitional cell mesotheliomas shed cells into the pleural fluid which have the combined (transitional) features of epithelial and sarcomatous mesothelioma cells (fig. 3-28).

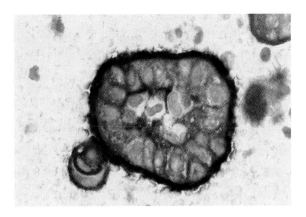

Figure 3-26

EPITHELIAL DIFFUSE MALIGNANT MESOTHELIOMA

Mesothelial cells in a pleural fluid cell block display strong membranous staining for epithelial membrane antigen. (Courtesy of Dr. R. Laucirica, Houston, TX.)

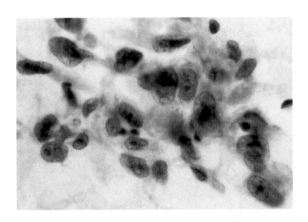

Figure 3-27

SARCOMATOUS DIFFUSE MALIGNANT MESOTHELIOMA

A fine-needle aspirate shows malignant spindle cells.

DIFFERENTIAL DIAGNOSIS

Cytologic Features of Carcinomas and Other Malignancies

Cytologic features common to most malignancies have been discussed previously and include increased cellularity, enlarged cells with enlarged nuclei (high nucleus to cytoplasm ratio), coarse chromatin, enlarged and multiple nucleoli, irregular or indented nuclear contours, increased mitoses, atypical mitoses, tumor necrosis, and characteristic three-dimensional architectural structure. The cytologic or

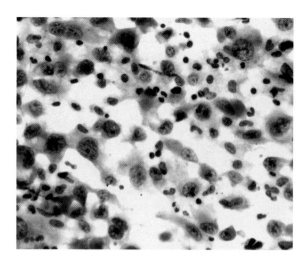

Figure 3-28

TRANSITIONAL DIFFUSE MALIGNANT MESOTHELIOMA

The cells have combined epithelial and sarcomatous features. (Courtesy of Dr. T. Allen, Tyler, TX.)

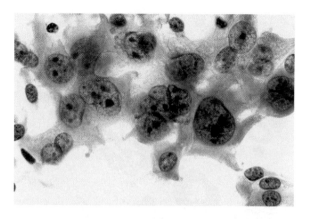

Figure 3-30

LUNG ADENOCARCINOMA

Cells with large irregular nuclei and multiple, large, red nucleoli are seen in pleural fluid. (Courtesy of Dr. M. L. Ostrowski, Houston, TX.)

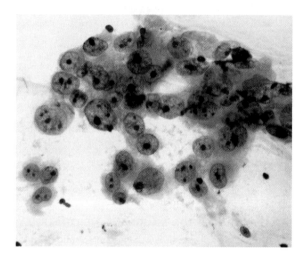

Figure 3-29

LUNG ADENOCARCINOMA

Cells in pleural fluid show enlarged nuclei with nuclear lobulation and prominent, often multiple, nucleoli.

architectural features of some metastatic cancers are characteristic for their primary site. Although many cancers still lack specific markers for their primary sites, immunostains of malignant effusions may be useful in identifying the primary site of a number of metastatic cancers or, at least, in limiting the differential diagnosis (1,3,6,7,10,14,15–21,24,27,28,33,36,37,39–41, 43–45,48,49,54,55,56,61,62,64,65).

Lung Carcinomas. Carcinoma of the lung is the most frequent cause of malignant pleural effusion, accounting for about 30 percent of the total (32). Adenocarcinomas cause the most malignant pleural effusions (3,49). The cells of adenocarcinoma of the lung tend to be large, with pale chromatin, thickened nuclear membranes, enlarged nucleoli often arranged in cohesive clusters, papillae or acini, and often, cytoplasmic mucin vacuoles (figs. 3-29–3-31). Well-differentiated peripheral adenocarcinomas invading the pleura consist of relatively homogeneous columnar cells with enlarged regular nuclei and prominent nucleoli.

The cells of large cell carcinoma of the lung are large and pleomorphic, and are observed as individual malignant cells or in clusters. Obvious cytologic features of malignancy include enlarged nuclei, coarse chromatin, and large, prominent nucleoli (3,49).

Both squamous cell carcinoma and small cell carcinoma are less likely than adenocarcinoma or large cell carcinoma to be found in pleural effusions (3,49). The cells of better differentiated squamous cell carcinoma have malignant cytologic features with dense cytoplasm and, in some cases, diagnostic keratin formation (fig. 3-32). The cells of poorly differentiated squamous cell carcinoma may be difficult to differentiate from cells of adenocarcinoma and may have cytoplasmic vacuoles.

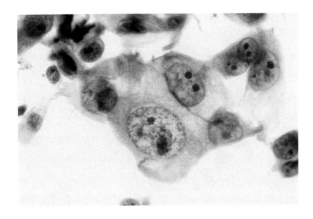

Figure 3-31

LUNG ADENOCARCINOMA

Cells in pleural fluid display large irregular nuclei with multiple large nucleoli. (Courtesy of Dr. M. L. Ostrowski, Houston, TX.)

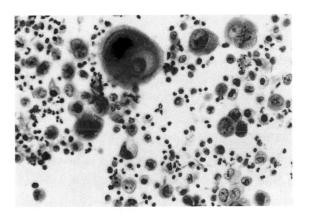

Figure 3-32

SQUAMOUS CELL CARCINOMA

The pleural fluid shows malignant cells with dense cytoplasm and irregular nuclei. (Courtesy of Dr. R. Laucirica, Houston, TX.)

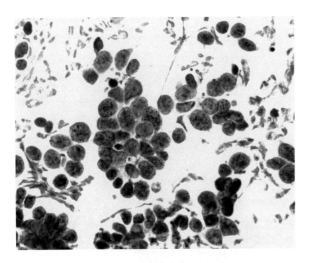

Figure 3-33

SMALL CELL CARCINOMA

The cells in this bloody pleural effusion show the characteristic features of finely stippled chromatin, scant cytoplasm, and nuclear molding.

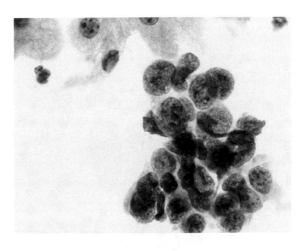

Figure 3-34

SMALL CELL CARCINOMA

Cancer cells in a pleural effusion have scant cytoplasm, finely stippled chromatin, and nuclear molding. Some cells have visible nucleoli. (Courtesy of Dr. M. L. Ostrowski, Houston, TX.)

The cells of small cell carcinoma are hyperchromatic, with scant cytoplasm and finely stippled ("salt and pepper") chromatin arranged in chains or clusters (2,3,49). The cells may show nuclear molding, with adjacent cells fitting together somewhat like pieces of a puzzle (figs. 3-33–3-35). Although the name may be misleading, most cells of small cell carcinoma are larger than lymphocytes. On occasion, the better-preserved small cell carcinoma cells in effusion fluids have more abundant cytoplasm and the presence of nucleoli, and can be mistaken for nonsmall cell carcinoma.

Breast Carcinomas. Metastatic breast carcinoma causes about 25 percent of malignant pleural effusions (29). Breast cancer cells are characterized by irregular nuclei and multiple nucleoli, but are generally not as pleomorphic as cells from many carcinomas from other primary sites (3,49). The cells are classically observed in abundant

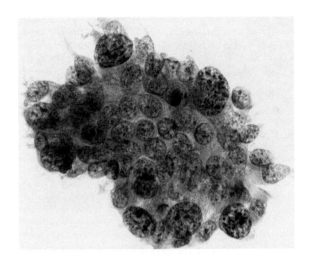

Figure 3-35

NEUROENDOCRINE CARCINOMA

The cancer cells have the stippled chromatin characteristic of neuroendocrine carcinomas. (Courtesy of Dr. M. L. Ostrowski, Houston, TX.)

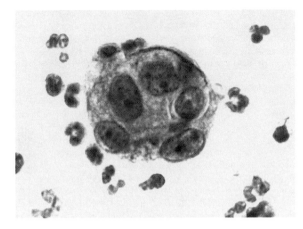

Figure 3-36

BREAST CARCINOMA

A metastasis in pleural fluid shows an acinar structure, with cells in a gland-like arrangement. Compare the size of the cancer cells to adjacent neutrophils.

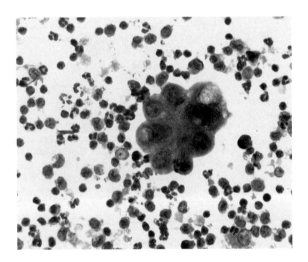

Figure 3-37

BREAST CARCINOMA

Metastatic cancer cells in pleural fluid have an acinar arrangement and several show cytoplasmic vacuoles. The relatively smooth border of the cell group is common in breast cancer cells found in fluids. The tumor cells dwarf the surrounding lymphocytes and neutrophils.

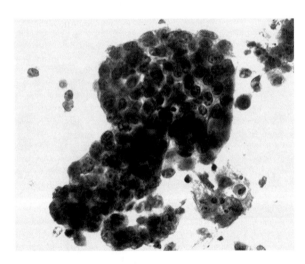

Figure 3-38

BREAST CARCINOMA

The large metastatic cancer cell balls form a three-dimensional papillary structure. This is characteristic of breast cancer in pleural fluid.

three-dimensional morulae but can be in papillae or as individual cells (figs 3-36–3-40).

Lobular carcinoma of the breast forms chains of small, homogeneous, hyperchromatic cells, sometimes with the nucleus compressed by a large cytoplasmic vacuole, forming signet ring cells. The nuclei of lobular carcinoma can be quite bland but usually have small nucleoli.

Gastrointestinal Carcinomas. In effusions, the cells of gastrointestinal carcinomas have frankly malignant cytologic features and often contain cytoplasmic mucin (3,49). The nuclei of these cells may be compressed by a large cytoplasmic mucin vacuole, forming a signet ring

Cytology of the Serosal Surfaces

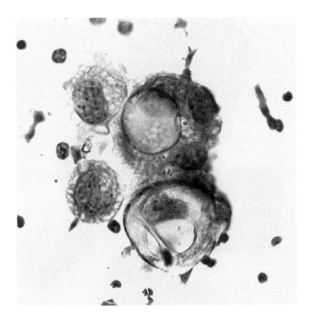

Figure 3-39

BREAST CARCINOMA

Metastatic breast cancer cells in a bloody pleural effusion display cytoplasmic vacuoles.

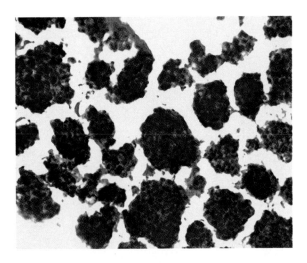

Figure 3-40

BREAST CARCINOMA

Multiple morulae of metastatic breast carcinoma are seen in a pleural fluid cytology specimen.

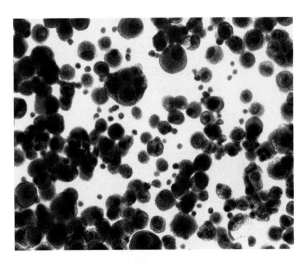

Figure 3-41

GASTROINTESTINAL CARCINOMA

Metastatic adenocarcinoma cells in a pleural effusion consist of small clusters and individual cells. The consistent eccentric position of the nucleus in these tumor cells suggests gastric carcinoma as the primary.

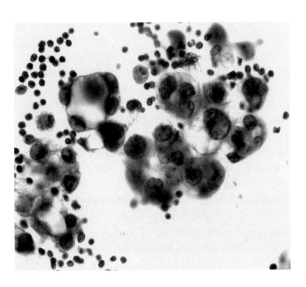

Figure 3-42

GASTROINTESTINAL CARCINOMA

Clusters of metastatic adenocarcinoma cells in a pleural effusion include many cells with cytoplasmic vacuoles.

cell (figs. 3-41–3-43). Although gastric cancers are the classic source of signet ring cells in malignant effusions, adenocarcinomas from other sites also produce such cells. Some mucinous adenocarcinomas display a few clusters of relatively bland mucin-containing cells floating in a background of basophilic mucin.

Other Carcinomas. Virtually any carcinoma can metastasize to the pleura, peritoneum, or pericardium and create a malignant effusion (3,12,49). On occasion, the cytologic detection

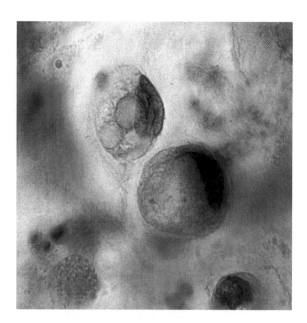

Figure 3-43

GASTROINTESTINAL CARCINOMA

The pleural fluid contains cells with mucin-filled cytoplasmic vacuoles.

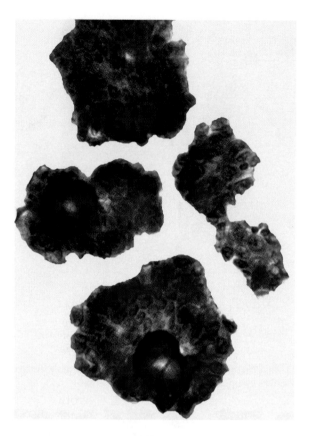

Figure 3-44

OVARIAN CARCINOMA

Morulae of metastatic ovarian carcinoma with psammoma bodies.

of metastatic carcinoma in pleural or peritoneal fluid provides the initial diagnosis of a patient's carcinoma (fig. 3-44). Whether or not the effusion contains metastatic cancer cells depends in part on the type of cancer. Serous papillary carcinomas may be primary in the peritoneum and shed cells into the ascitic fluid (figs. 3-45, 3-46). As with sarcomatous mesotheliomas, cells of sarcomas and sarcomatous carcinomas are less likely to exfoliate into effusion fluid. Sarcomas involving the serosal membranes may be diagnosed by fine-needle aspiration cytology (fig. 3-47). Lymphomatous effusions typically contain lymphoma cells (fig. 3-48). Melanomas can metastasize to the pleura and the characteristic cells may be found in effusion fluid (figs. 3-49, 3-50).

Cytologic Features of Non-Neoplastic Effusions

The cytologic findings in some specific non-neoplastic effusions are mentioned here because non-neoplastic effusions are frequently encountered and, when hyperplastic mesothelial cells with reactive atypia are present, enter into the differential diagnosis of diffuse malignant mesothelioma and other malignancies.

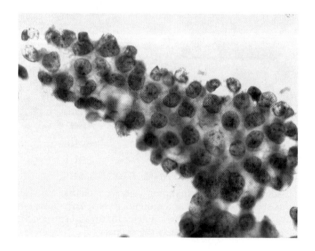

Figure 3-45

PAPILLARY SEROUS CARCINOMA OF THE PERITONEUM

The three-dimensional papilla consists of overlapping pleomorphic cells in the peritoneal fluid.

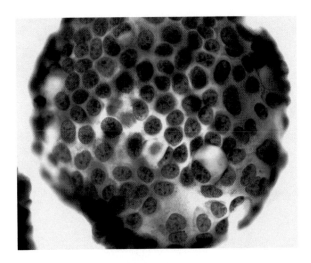

Figure 3-46

**PAPILLARY SEROUS
CARCINOMA OF THE PERITONEUM**

The three-dimensional ball-shaped morula shows overlapping cells. The flat and smooth outer border of the cell group is characteristic and mimics breast cancer.

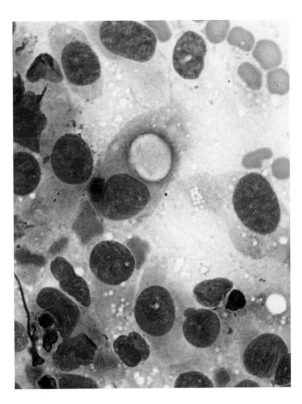

Figure 3-47

EPITHELIOID HEMANGIOENDOTHELIOMA

Cytology specimen of epithelioid hemangioendothelioma shows a tumor cell with a cytoplasmic lumen. (Courtesy of Dr. M. L. Ostrowski, Houston, TX.)

Blood, inflammatory cells, and necrotic debris can be observed in cytology specimens from both benign and malignant effusions (3,34,49).

Infections and Empyemas. Effusions due to bacterial infections and empyemas are characterized by abundant neutrophils (3,49). In addition to pleural infections, intra-abdominal abscesses also cause exudative pleural effusion with neutrophils. Eosinophils are less often seen in infectious pleural effusions caused by parasites, *Mycobacteria tuberculosis*, or fungi.

Pleural effusions occur in about one third of patients with tuberculosis. They are characterized by abundant lymphocytes, which are thought to be the result of an immunologic reaction. Few organisms are identified (15,30,31,50). The sensitivity of various diagnostic techniques for analyzing tuberculosis effusions are: 6 percent for acid-fast staining, 17 percent for cultures, and 50 percent for polymerase chain reaction (PCR) amplification (50). Necrosis, a feature often associated with malignancy, can be seen in infectious effusions, especially with tuberculosis.

Underlying Pulmonary Diseases. A unilateral, bloody pleural effusion can be caused by cancers but also by pulmonary infarcts. The reactive mesothelial cells that shed into the bloody effusion fluid overlying an infarct can

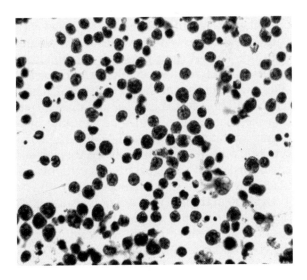

Figure 3-48

LYMPHOMA

Large numbers of lymphoma cells are seen in the pleural effusion. Nuclear clefting, an open chromatin pattern, and visible nucleoli raise the suspicion for lymphoma.

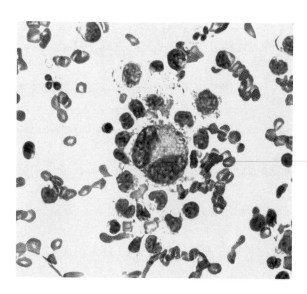

Figure 3-49

MELANOMA

Binucleated melanoma cell in bloody pleural effusion.

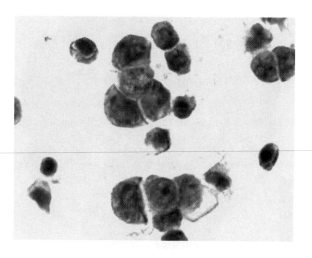

Figure 3-50

MELANOMA

Metastatic melanoma in pleural fluid consists of cells with irregular nuclei, nucleoli, and nuclear molding. Pigment is lacking.

be notoriously atypical, potentially causing confusion with a malignant effusion (51). Other underlying pulmonary diseases, including pneumonias and lung cancers, can cause effusions that contain benign hyperplastic mesothelial cells with reactive atypia. Graft rejection in lung transplant patients causes pleural effusions that contain abundant lymphocytes.

Collagen Vascular Diseases. Collagen vascular disease is a well-known cause of pleural effusion. The effusion caused by rheumatoid arthritis is characterized cytologically by histiocytes, neutrophils, and necrotic debris. Characteristic multinucleated and spindle-shaped cells of a rheumatoid nodule may be seen in a fine-needle aspiration specimen (3,47,49). The characteristic cells of systemic lupus erythematosus are formed when a phagocytized neutrophil displaces the nucleus of a leukocyte (3,38,49).

Drug Reactions. Drug reactions cause effusions that are typified by the presence of pleural eosinophilia, defined as effusion fluid containing eosinophils representing more than 10 percent of the exfoliated cells (3,29,49). Chemotherapy may result in the presence of atypical reactive mesothelial cells and even necrosis.

Pneumothorax/Hemothorax. The most common causes of pleural fluid eosinophilia are pneumothorax and hemothorax (29,32). Hemothorax is defined by a pleural effusion hematocrit of more than 50 percent of the patient's peripheral blood hematocrit. Red blood cells and hemosiderin are seen in the pleural fluid. Aggregates of hemosiderin particles are referred to as Heinz bodies.

Other Causes. There are numerous other causes of effusions and many of these, including those associated with heart failure and cirrhosis, are associated with reactive atypical mesothelial cells on exfoliative cytology (3,9,29, 32,49,66). Endometriosis may be seen in peritoneal fluids and is, rarely, associated with pleural effusion. Cell aggregates of endometriosis should not be confused with malignant morulae. Asbestos exposure can produce benign pleural effusions and, therefore, a specific diagnosis of malignant pleural effusion should not be based merely on a history of asbestos exposure. Some pleural effusions remain nonspecific or idiopathic even after extensive workup and a diagnosis is never obtained in 10 to 20 percent of cases.

REFERENCES

1. Ascoli V, Scalzo CC, Taccogna S, Nardi F. The diagnostic value of thrombomodulin immunolocalization in serous effusions. Arch Pathol Lab Med 1995;119:1136–40.

2. Banner BF, Warren WH, Gould VE. Cytomorphology and marker expression of malignant neuroendocrine cells in pleural effusions. Acta Cytol 1986;30:99–104.

3. Battifora H. The pleura. In: Sternberg SS, ed. Diagnostic surgical pathology, 2nd ed. New York: Raven Press; 1994:1095–123.

4. Beaty MW, Fetsch P, Wilder AM, Marincola F, Abati A. Effusion cytology of malignant melanoma. A morphologic and immunocytochemical analysis including application of the MART-1 antibody. Cancer (Cancer Cytopathol) 1997; 81:57–63.

5. Berge T, Grontoft O. Cytologic diagnosis of malignant pleural mesothelioma. Acta Cytol 1965;9:207–12.

6. Boon M, Veldhuizen RW, Ruinaard C, Snieders MW, Kwee WS. Qualitative distinctive differences between the vacuoles of mesothelioma cells and of cells from metastatic carcinoma exfoliated in pleural fluid. Acta Cytol 1984; 28:443–9.

7. Bramwell ME, Ghosh AK, Smith WD, Wiseman G, Spriggs AI, Harris H. Ca2 and Ca3. New monoclonal antibodies evaluated as tumor markers in serous effusions. Cancer 1985;56:105–10.

8. Carrillo R, Sneige N, el-Naggar AK. Interphase nucleolar organizer regions in the evaluation of serosal cavity effusions. Acta Cytol 1994;38: 367–72.

9. Casarett GW. Radiation histopathology. Boca Raton, FL: CRC Press; 1980.

10. Chhieng DC, Yee H, Schaefer D, et al. Calretinin staining pattern aids in the differentiation of mesothelioma from adenocarcinoma in serous effusions. Cancer 2000;90:194–200.

11. Choi YL, Song SY. Cytologic clue of so-called nodular histiocytic hyperplasia of the pleura. Diagn Cytopathol 2001;24:256–9.

12. Cuijpers VM, Boerman OC, Salet van de Pol MR, Vooijs GP, Poels LG, Ramaekers FC. Immunocytochemical detection of ovarian carcinoma cells in serous effusions. Acta Cytol 1993;37:272–9.

13. DiBonito L, Falconieri G, Colautti I, Bonifacio D, Dudine S. The positive pleural effusion. A retrospective study of cytopathologic diagnosis with autopsy confirmation. Acta Cytol 1992; 36:329–32.

14. Esteban JM, Yokota S, Husain S, Battifora H. Immunocytochemical profile of benign and carcinomatous effusions. A practical approach to difficult diagnosis. Am J Clin Pathol 1990;94:698–705.

15. Ferrer J. Pleural tuberculosis. Eur Respir J 1997;10:942–7.

16. Fetsch PA, Abati A. Immunocytochemistry in effusion cytology: a contemporary review. Cancer 2001;93:293–308.

17. Fetsch PA, Abati A, Hijazi YM. Utility of the antibodies CA 19-9, HBME-1 and thrombomodulin in the diagnosis of malignant mesothelioma and adenocarcinoma in cytology. Cancer 1998;84:101–8.

18. Fetsch PA, Simsir A, Abati A. Comparison of antibodies to HBME-1 and calretinin for the detection of mesothelial cells in effusion cytology. Diagn Cytolpathol 2001;25:158–61.

19. Ghosh AK, Spriggs AI, Taylor-Papadimitriou J, Mason DY. Immunocytochemical staining of cells in pleural and peritoneal effusions with a panel of monoclonal antibodies. J Clin Pathol 1983;36:1154–64.

20. Gioanni J, Caldani C, Zanghellini E, et al. A new epithelial membrane antigen (Calam 27) as a marker of carcinoma in serous effusions. Acta Cytol 1991;35:315–9.

21. Hilborne LH, Cheng L, Nieberg RK, Lewin KJ. Evaluation of an antibody to human milk fat globule antigen in the detection of metastatic carcinoma in pleural, pericardial and peritoneal fluids. Acta Cytol 1986;30:245–50.

22. Huang MS, Tsai MS, Hwang JJ, Wang TH. Comparison of nucleolar organiser regions and DNA flow cytometry in the evaluation of pleural effusion. Thorax 1994;49:1152–6.

23. Kho-Duffin J, Tao LC, Cramer H, Catellier MJ, Irons D, Ng P. Cytologic diagnosis of malignant mesothelioma, with particular emphasis on the epithelial noncohesive cell type. Diagn Cytopathol 1999;20:57–62.

24. Khoor A, Whitsett JA, Stahlman MT, Olson SJ, Cagle PT. Utility of surfactant protein B precursor and thyroid transcription factor 1 in differentiating adenocarcinoma of the lung from malignant mesothelioma. Hum Pathol 1999;30: 695–700.

25. Klempman S. The exfoliative cytology of diffuse pleural mesothelioma. Cancer 1962;15: 691–704.

26. Lee A, Baloch ZW, Yu G, Gupta PK. Mesothelial hyperplasia with reactive atypia: diagnostic pitfalls and role of immunohistochemical studies—a case report. Diagn Cytopathol 2000;22:113–6.

27. Leong AS. Immunostaining of cytologic specimens. Am J Clin Pathol 1996;105:139–40.

28. Li CY, Lazcano-Villareal O, Pierre RV, Yam LY. Immunocytochemical identification of cells in serous effusions. Technical considerations. Am J Clin Pathol 1987;88:696–706.

29. Light RW. Clinical diagnosis of pleural disease. In: Cagle PT, ed. Diagnostic pulmonary pathology. New York: Marcel Dekker; 2000:571–81.

30. Light RW. Establishing the diagnosis of tuberculous pleuritis. Arch Intern Med 1998;158:1967–8.

31. Light RW. Tuberculous pleural effusions. In: Light RW, ed. Pleural diseases, 3rd ed. Baltimore: Williams & Wilkins; 1995:154–6.

32. Light RW, MacGregor MI, Luchsinger PC, Ball WC Jr. Pleural effusions: the diagnostic separation of transudates and exudates. Ann Intern Med 1972;77:507–13.

33. Lozano MD, Panizo A, Toledo GR, Sola JJ, Pardo-Mindan J. Immunocytochemistry in the differential diagnosis of serous effusions: a comparative evaluation of eight monoclonal antibodies in Papanicolaou stained smears. Cancer 2001;93:68–72.

34. Luse SA, Reagan J. A histocytological study of effusions. Effusions not associated with malignant tumors. Cancer 1954;7:1155–66.

35. Mallonee MM, Lin F, Hassanein R. A morphologic analysis of the cells of ductal carcinoma of the breast and of adenocarcinoma of the ovary in pleural and abdominal effusions. Acta Cytol 1987;31:441–7.

36. Martin SE, Moshiri S, Thor A, Vilasi V, Chu EW, Schlom J. Identification of adenocarcinoma in cytospin preparations of effusions using monoclonal antibody B72.3. Am J Clin Pathol 1986;86:10–8.

37. Mason MR, Bedrossian CW, Fahey CA. Value of immunocytochemistry in the study of malignant effusions. Diagn Cytopathol 1987;3:215–21.

38. Metzger AL, Coyne M, Lee S, Kramer LS. In vivo LE cell formation in peritonitis due to systemic lupus erythematosus. J Rheumatol 1974;1:130–3.

39. Miedouge M, Rouzaud P, Salama G, et al. Evaluation of seven tumour markers in pleural fluid for the diagnosis of malignant effusions. Br J Cancer 1999;81:1059–65.

40. Monte S, Ehya H, Lang WR. Positive effusion cytology as the initial presentation of malignancy. Acta Cytol 1987;31:448–52.

41. Mu XC, Brien TP, Ross JS, Lowry CV, McKenna BJ. Telomerase activity in benign and malignant cytologic fluids. Cancer 1999;87:93–9.

42. Mullick SS, Green LK, Ramzy I, et al. p53 gene product in pleural effusions. Practical use in distinguishing benign from malignant cells. Acta Cytol 1996;40:855–60.

43. Murphy W, Ng AB. Determination of primary site by examination of cancer cells in body fluids. Am J Clin Pathol 1972;58:479–88.

44. Nagel H, Hemmerlein B, Ruschenburg I, Huppe K, Droese M. The value of anti-calretinin antibody in the differential diagnosis of normal and reactive mesothelia versus metastatic tumors in effusion cytology. Pathol Res Pract 1998;194:759–64.

45. Ng WK, Chow JC, Ng PK. Thyroid transcription factor-1 is highly sensitive and specific in differentiating metastatic pulmonary from extrapulmonary adenocarcinoma in effusion fluid cytology specimens. Cancer 2002;96:43–8.

46. Pedio G, Landolt-Weber U. Cytologic presentation of malignant mesothelioma in pleural effusion. Exp Cell Biol 1988;56:211–6.

47. Petty TL, Wilkins M. The five manifestations of rheumatoid lung. Dis Chest 1996;49:75–82.

48. Pinto MM. An immunoperoxidase study of S-100 protein in neoplastic cells in serous effusions. Use as a marker for melanoma. Acta Cytol 1986;30:240–4.

49. Ramzy I. Clinical cytopathology and aspiration biopsy, 2nd ed. New York: McGraw Hill; 2001.

50. Reechaipichitkul W, Lulitanond V, Sungkeeree S, Patjanasoontorn B. Rapid diagnosis of tuberculous pleural effusion using polymerase chain reaction. Southeast Asian J Trop Med Public Health 2000;31:509–14.

51. Romero Candeira S, Hernandez Blasco L, Soler MJ, Munoz A, Aranda I. Biochemical and cytologic characteristics of pleural effusions secondary to pulmonary embolism. Chest 2002;121:465–9.

52. Sears D, Hajdu S. The cytologic diagnosis of malignant neoplasms in pleural and peritoneal effusions. Acta Cytol 1995;31:85–97.

53. Sherman M, Mark EJ. Effusion cytology in the diagnosis of malignant epithelioid and biphasic pleural mesothelioma. Arch Pathol Lab Med 1990;114:845–51.

54. Silverman JF, Nance K, Phillips B, Norris HT. The use of immunoperoxidase panels for cytologic diagnosis of malignancy in serous effusions. Diagn Cytopathol 1987;3:134–40.

55. Simsir A, Fetsch PA, Mehta D, Zakowski M, Abati A. E-cadherin, N-cadherin and calretinin in pleural effusions: the good, the bad, the worthless. Diagn Cytopathol 1999;20:125–30.

56. Stevens MW, Leong AS, Fazzalari NL, Dowling KD, Henderson DW. Cytopathology of malignant mesothelioma: a stepwise logistic regression analysis. Diagn Cytopathol 1992;8:333–41.

57. Suthipintawong C, Leong AS, Vinyuvat S. Immunostaining of cell preparations: a comparative evaluation of common fixatives and protocols. Diagn Cytolopathol 1996;15:167–74.

58. Tao LC. The cytopathology of mesothelioma. Acta Cytol 1979;23:209–13.

59. Tiniakos DG, Healicon RM, Hair T, Wadehra V, Home CH, Angus B. p53 immunostaining as a marker of malignancy in cytologic preparations of body fluids. Acta Cytol 1995;39:171–6.

60. Triol JH, Conston AS, Chandler SV. Malignant mesothelioma. Cytopathology of 75 cases seen in a New Jersey community hospital. Acta Cytol 1984;28:37–45.

61. van Niekerk CC, Jap PH, Thomas CM, Smeets DF, Ramaekers FC, Poels LG. Marker profile of mesothelial cells versus ovarian carcinoma cells. Int J Cancer 1989;43:1065–71.

62. Walts AE, Said JW, Shintaku IP. Epithelial membrane antigen in the cytodiagnosis of effusions and aspirates: immunocytochemical and ultrastructural localization in benign and malignant cells. Diagn Cytopathol 1987;3:41–9.

63. Whitaker D, Shilkin KB. Diagnosis of pleural malignant mesothelioma in life—a practical approach. J Pathol 1984;143:147–75.

64. Wick MR, Moran CA, Mills SE, Suster S. Immunohistochemical differential diagnosis of pleural effusions, with emphasis on malignant mesothelioma. Curr Opin Pulm Med 2001;7:187–92.

65. Wieczorek TJ, Krane JF. Diagnostic utility of calretinin immunohistochemistry in cytologic cell block preparations. Cancer 2000;90:312–9.

66. Wojno KJ, Olson JL, Sherman ME. Cytopathology of pleural effusions after radiotherapy. Acta Cytol 1994;38:1–8.

67. Zoppi JA, Pellicer EM, Sundblad AS. Diagnostic value of p53 protein in the study of serous effusions. Acta Cytol 1995;39:721–4.

4 DIFFUSE MALIGNANT TUMORS OF THE SEROSAL MEMBRANES

DIFFUSE MALIGNANT MESOTHELIOMA

Diffuse malignant mesothelioma is a primary diffuse tumor of the serosal membranes that recapitulates many of the forms of normal and reactive mesothelial and submesothelial cells. In the past, before the mesothelial nature of these tumors was recognized, they were referred to by a variety of names, but *diffuse malignant mesothelioma* and the common working abbreviation, *malignant mesothelioma*, are the only currently acceptable terms for diagnosis.

EPIDEMIOLOGY

Incidence

Diffuse malignant mesothelioma is an uncommon tumor. The current incidence rate for men in the United States and Canada is about 20 cases/million persons/year, but is only 2 to 3 cases/million/year for women (fig. 4-1) (seer.cancer.gov). There has been a steady increase in the incidence rate for men over the last 30 to 40 years, whereas for women it has remained essentially unchanged over this period (fig. 4-1). These gender differences are believed to be due primarily to past asbestos exposure (see below). If the proportion of mesotheliomas believed to be associated with asbestos exposure is removed from the calculation, the incidence rate for men falls to 2 to 3 cases/million, suggesting that this is the background (nonasbestos-induced) mesothelioma incidence rate (159). Modeling of observed incidence rates (135) and actual observational data (fig. 4-1) indicate that the peak incidence of cases in the United States occurred in the early 1990s, and is now slowly declining. Much higher incidence rates are seen in Great Britain and Australia: for adult men in Australia in 1997 the rate was 60 and for adult women 11/million persons/year (87). These numbers appear to reflect the widespread use of crocidolite asbestos, a particularly potent inducer of mesothelioma. Case numbers in these two countries are

still increasing. High rates are also seen in some countries in Europe (83).

Location of Tumors

Mesotheliomas occur in the pleural, peritoneal, and pericardial cavities, as well as in the tunica vaginalis testis. Peritoneal tumors may also present in hernia sacs. The majority of tumors are pleural in origin, followed by peritoneal primaries, but again there are gender differences. For men, the ratio of pleural to peritoneal tumors in a recent study in the United States was 9 to 1, but for women only 2 to 1 (158). This difference probably reflects the additional tumors produced in men in the pleural cavity by asbestos exposure. In Britain, where women appear to have more exposure to asbestos, the ratio is about 5 to 1 (17). Pericardial and tunica vaginalis lesions are rare. The same histologic patterns are seen in all locations, although purely sarcomatous lesions are uncommon outside the pleural cavity.

ETIOLOGY

Asbestos

Table 4-1 shows a list of accepted and proposed causes of mesothelioma. By far the most extensively investigated cause is asbestos exposure. The association of asbestos exposure and mesothelioma was first suggested in 1960 by Wagner et

Table 4-1
CAUSES OF MESOTHELIOMA
Exposure to asbestos
Erionite exposure (in Turkey only)
Therapeutic radiation (disputed)
Chronic pleural irritation (rare)
Simian virus (SV)40
Idiopathic

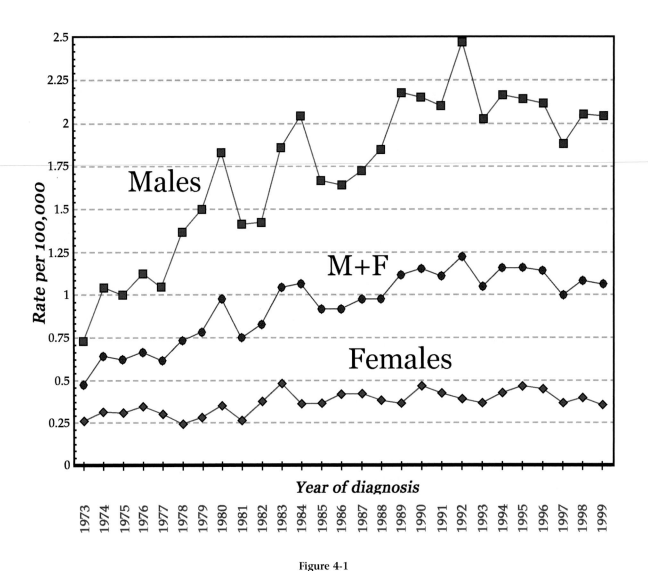

Figure 4-1

AGE-ADJUSTED INCIDENCE OF MESOTHELIOMA IN THE UNITED STATES FROM 1973 TO 1999

Data from the Surveillance Epidemiology and End Results registry (SEER).

al. (175). In the numerous subsequent investigations that have confirmed this observation, the exact proportion of patients giving a history of asbestos exposure varies considerably by series, and particularly by which occupational groups are studied. A recent report on patients from a set of cancer registries in the United States (159), however, which looked at the causes of mesothelioma in a large population not selected by prior knowledge of occupation, showed asbestos exposure to be the cause of 90 percent of the pleural tumors and about 60 percent of the peritoneal tumors in men. In women, literature re-

sults are more variable than for men, but the same study reported a history of asbestos exposure in only 23 percent of women. These numbers apply to the United States and may be different in other countries; however, they do illustrate the point that some proportion of mesotheliomas are not associated with asbestos exposure.

Historically, individuals in a variety of occupations have been at risk for developing mesothelioma (92,145). These include career asbestos workers (usually pipefitters or insulators), individuals who work in shipyards, individuals involved in asbestos cement production in plants where both amphibole and chrysotile (see below) asbestos are used, individuals working in asbestos textile or friction products plants where both amphibole and chrysotile are used, and amphibole asbestos miners. The mesothelioma risk for those in chrysotile mining operations and chrysotile only friction and cement plants has been low. A variety of workers in nonregulated industries, for example, electricians, carpenters, and general laborers involved in construction work, also have an increased risk. Pleural mesotheliomas are also seen as a result of household contact exposure, such as washing asbestos-contaminated workclothes.

Asbestos occurs in two mineralogic forms: commercially used amphiboles (amosite and crocidolite) and chrysotile. Amosite and crocidolite are considerably more persistent in tissue than is chrysotile, which disappears from lung tissue rapidly (27). This difference in biopersistence is believed to be the reason that mesothelioma incidence rates are much higher in those who used or manufactured products with amosite and crocidolite compared to only chrysotile (92). There is some debate about the fiber type–specific ratio of risk: a recent summary proposes that the relative risk is 1 for chrysotile, 100 for amosite, and 500 for crocidolite (64). Other estimates with lower ratios exist but the differences are still marked, and some cohorts with exposure only to chrysotile have not shown increases in mesothelioma incidence (92).

The latency period (time from first exposure to the appearance of disease) for the development of mesothelioma after asbestos exposure is generally very long, with a mean of 30 to 40 years (153). Latency periods of less than 20 years are uncommon and mesotheliomas rarely, if ever, occur less than 15 years after first exposure (84).

Other Causes of Mesothelioma

A small number of cases of mesothelioma have been reported in persons who received therapeutic radiation as treatment for a malignant neoplasm (23), although a recent large epidemiologic study (109) examining individuals who received thoracic radiation for breast cancer or Hodgkin's disease failed to find any increased incidence of mesothelioma. A few cases have been reported after plombage therapy for tuberculosis or tuberculous empyema (61,108,

147). Other causes of chronic irritation of the pleura may also produce mesothelioma (61).

In Turkey there is a localized region where the soil and volcanic outcrop rock contain an asbestos-sized (but not asbestos) biopersistent fiber called erionite. This fiber is a very potent cause of mesothelioma in the villages in this region (11), but cases of erionite-induced mesothelioma have not been reported elsewhere.

Recently it has been proposed that Simian virus 40 (SV40) contamination of polio vaccines in the 1950s might have lead to widespread infection of the population and a subsequent role for SV40 in the development of human mesotheliomas. SV40 does produce mesotheliomas in some species of experimental animal; however, epidemiologic analysis of human populations have not shown consistent or significant associations of mesotheliomas with exposure of individuals to potentially contaminated polio vaccines, and a recent large study has specifically denied such an association (163). At this point the relationship between SV40 and human mesothelioma, if any, is obscure.

As indicated above, mesotheliomas occur in persons who have had asbestos exposure and those who have not, and a history of exposure to asbestos or lack of exposure is important in assigning causation to a given tumor. A history of exposure to asbestos should play no role in diagnosis, however; diagnosis depends only on the gross, microscopic, and special technique observations, as it does with any other tumor.

PATHOGENESIS

Most investigators believe that mesotheliomas arise from mesothelial cells, although multipotential subserosal mesenchymal cells have also been suggested as cells of origin (14,15). Studies in laboratory animals and observations of human tissues suggest that there are histopathologic stages in the development of mesothelioma similar to the better established early stages of other malignancies: from mesothelial hyperplasia, to dysplasia, to in situ mesothelioma, to minimal invasion (36,58,182). In animal studies, instillation of asbestos or similar fibers in a body cavity results in inflammation of subserosal tissues and mesothelial hyperplasia, which in some cases progresses to dysplasia, with some of those progressing to mesothelioma (47,48,150,174). As opposed to other cancers, such as lung cancer or cervical cancer, in which the precursor lesions are fairly well defined, there is little detailed information about the exact morphologic and molecular changes that precede the appearance of overt mesothelioma.

The term *mesothelioma in situ* has been used for the changes seen in the surface mesothelium in human tissues adjacent to foci of frank invasion (6,58,182). Henderson et al. (58,182) proposed that in situ mesothelioma featured a noninvasive surface mesothelium with 1) abnormal architecture (linear, papillary, tubulopapillary), 2) cytologic atypia, and 3) absence of a background of exudative inflammation. They emphasized, however, that a diagnosis of mesothelioma in situ could only be made if invasion is identified in a different area or at a different time, and there is no proof that the surface lesions they observed really represented mesothelioma in situ. At the present time, there are no universally accepted morphologic criteria for the diagnosis of mesothelioma in situ (see chapter 5). In addition to observations in asbestos-exposed individuals, mesothelial papillary proliferation, which was proposed to be a precursor to mesothelioma, has been reported in a patient subsequent to radiotherapy for lung cancer (66); again, there is no proof that these lesions were mesothelioma in situ.

Steps in the molecular pathogenesis of diffuse malignant mesothelioma do not follow the typical patterns seen with many other solid cancers, and mutations of tumor suppressor genes and oncogenes common to many malignancies have rarely been identified in mesotheliomas (21,79,110). Most genetic changes described in mesotheliomas are chromosomal deletions, and the mechanisms by which asbestos, traditionally the cause of most mesotheliomas, contributes to their development are still not well understood. Diffuse malignant mesotheliomas have many chromosomal aberrations when analyzed by karyotypic, DNA cytometric, and comparative genomic hybridization studies (39,81,106). Although mesotheliomas have a fairly typical pattern of cytogenetic defects, there is no one chromosomal abnormality that is specific.

In contrast to many other types of solid tumor, diffuse malignant mesotheliomas generally lack mutations of the fundamental tumor suppressor genes, *p53* and *RB* (21,79,110). Only a few molecular defects have been identified in malignant mesothelioma thus far. In a recent study, 31 percent of diffuse malignant mesotheliomas had *p16* mutations (63). Mutations have also been reported in *p14ARF* and *NF2* (25,129). Mutations of the Wilms' tumor gene (*WT1*) are found in only a small percentage of mesotheliomas despite the utility of *WT1* expression as a mesothelioma marker (151). Detection of DNA and proteins consistent with the T-antigen of the SV40 virus has led to the hypothesis that this virus may have some role in the development of diffuse malignant mesothelioma. Some investigators have suggested that the ability of the SV40 virus to bind to p53 and/or RB products and inhibit them, without causing mutation, may explain the development of malignancy even though mutations in *p53* and *RB* are rare in diffuse malignant mesothelioma (169). It has also been reported that the SV40 virus can induce telomerase activity in human mesothelial cells (46).

CLINICAL FEATURES

The median age for presentation with malignant mesothelioma is around 60 years; mesothelioma is uncommon in men under age 50. Women with peritoneal mesotheliomas have a wide age range, with a much larger proportion seen in young women (76). Mesotheliomas occasionally occur in children and teenagers, and are clinically and morphologically identical to those in older individuals (52,86).

Patients with pleural tumors commonly present with chest pain and shortness of breath, sometimes accompanied by weight loss, cough, or fever. On rare occasions, there are no symptoms and the tumor is discovered through the incidental observation of a pleural effusion or pleural thickening on chest X ray. A few patients have presented with pneumothorax (156), metastases to cervical nodes (165), and even multiple small intrapulmonary metastases mimicking miliary tuberculosis (TB) on chest X ray in the absence of obvious pleural disease (107). As a general rule, however, patients who present with metastatic disease are more likely to have an underlying carcinoma, particularly a lung cancer, than a mesothelioma.

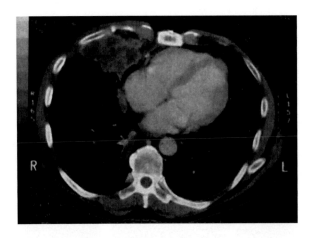

Figure 4-2

MALIGNANT MESOTHELIOMA

Computerized tomography (CT) scan of a pleural malignant mesothelioma shows tumor surrounding lung and extending along an interlobular fissure.

Radiographic examination shows a pleural effusion at the time of presentation in at least 90 percent of cases (12,94), and drainage of the effusion usually reveals either a diffuse pleural tumor or multiple pleural tumor nodules (fig. 4-2). In patients who have a history of asbestos exposure, pleural plaques or asbestosis may be visible radiographically. In some instances, tumor appears as mediastinal masses and occasionally as fairly circumscribed masses that may be difficult to distinguish from primary lung cancer (1,41,42,86).

The pleural effusion is commonly blood stained or overtly hemorrhagic (42,86). In many cases it is massive and tends to rapidly reaccumulate after pleurodesis. With progression of the disease, greater and greater amounts of tumor become radiographically visible, and the affected thorax sometimes becomes contracted. Thrombocythemia is a common finding. Clubbing, hypertrophic osteoarthropathy, hypoglycemia, and secretion of antidiuretic hormone (133), human chorionic gonadotropin (HCG) (140), and interleukin (IL)-6 (60) have been reported infrequently.

Patients with peritoneal tumors may present with abdominal pain, gastrointestinal complaints, ascites, and localized abdominal masses, including ovarian masses (33,103). Bowel obstruction is fairly common in advanced disease. On occasion, peritoneal tumors present in inguinal or umbilical hernias (111), and in very rare instances, as distant metastases (165). Computerized tomography (CT) typically reveals ascites, with diffuse omental or mesenteric thickening, or multiple small or large tumor masses (94, 146). In patients with a history of asbestos exposure, pleural plaques or asbestosis may be seen. Asbestosis is much more common in patients with peritoneal, rather than pleural, mesotheliomas, because induction of peritoneal mesotheliomas tends to require higher levels of asbestos exposure (64,176).

Pericardial mesotheliomas are rare: a review in 1994 (170) found only 27 cases in the English literature between 1972 and 1992. Patients with pericardial mesothelioma show a variety of fairly nonspecific signs and symptoms, including dyspnea, cardiac tamponade (173), constrictive pericarditis (181), arrhythmias, myocardial invasion and cardiac failure, and intracardiac thrombus formation (101).

Patients with mesotheliomas of the tunica vaginalis testis generally have scrotal enlargement, which is usually thought clinically to be either a hydrocele or a testicular tumor (16,134).

PATHOLOGIC FEATURES

Diagnosing malignant mesothelioma presents two problems. First is determining whether an obviously malignant tumor in the serosal membranes is a mesothelioma or is metastatic disease/direct spread of tumor from a nearby organ such as the lung or chest wall. Second is separating benign from malignant mesothelial proliferations. This chapter addresses the first issue; the second is considered in chapter 5.

Gross Features

By definition, malignant mesotheliomas are diffuse tumors that grow on a serosal surface and do not originate from the underlying organ. "Diffuse," in this setting, can mean multiple small tumor nodules (fig. 4-3), the typical appearance in fairly early disease; plaque-like masses of tumor; or large confluent sheets that form a rind completely or nearly completely surrounding the lung, abdominal viscera, or heart (see figs. 4-4, 4-6, 4-9). The rind may be several centimeters thick. Encasement by a rind of tumor is usually a sign of advanced disease,

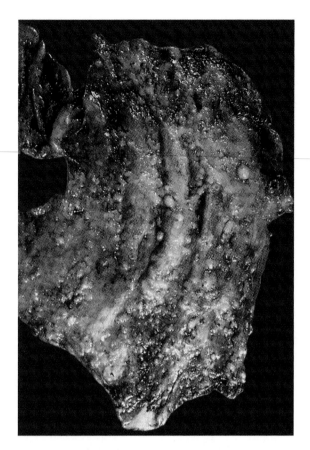

Figure 4-3

MALIGNANT MESOTHELIOMA

Malignant pleural mesothelioma appears as multiple small tumor nodules. (Fig. 4-7 from Fascicle 15, 3rd Series.)

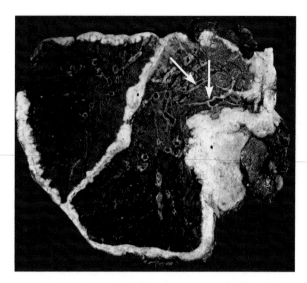

Figure 4-4

MALIGNANT MESOTHELIOMA

The tumor almost completely encases the lung and extends along the major fissure. Tumor also grows into the lung along the interlobular septa, surrounding small vessels and airways (arrows).

and with time this process may obliterate the pleural, pericardial, or peritoneal cavity. The presence of multiple tumor nodules or a rind over the serosal membranes is not specific for mesothelioma. Metastatic disease or other primary pleural malignancies may appear grossly identical (see Differential Diagnosis, below). In the pleural cavity, peripheral primary lung cancers or metastatic tumors also grow around the lung in a fashion that is a perfect mimic of a mesothelioma, and in the peritoneal cavity metastatic disease or primary papillary serous carcinoma can also mimic a mesothelioma.

The gross findings at surgical exploration are important for the pathologic diagnosis during life. Some information that indicates the presence of diffuse tumor should be sought before rendering a diagnosis. Radiographic reports may provide this information. The gross finding of

localized rather than diffuse disease should make one hesitate before diagnosing a diffuse malignant mesothelioma, although sometimes diffuse microscopic disease is present that is not apparent grossly. Similarly, the presence of disease thought to be benign on gross examination indicates a need for caution, although very early mesotheliomas that are only apparent microscopically do, rarely, occur (12,75).

Occasionally, pleural malignant mesotheliomas form large bulky masses that can mimic primary lung cancers, but there is usually obvious tumor elsewhere along the serosal membranes. Some epithelial mesotheliomas produce copious amounts of hyaluronate; these tumors may contain grossly visible cysts filled with thick, slightly yellow, mucoid material. Similar material may fill the remaining pleural cavity.

In the pleural cavity, tumor frequently extends along the fissures between the lobes of the lung and also grows into the lung along the interlobular septa (fig. 4-4). Tumor spreading in this fashion can reach the airspaces or enter the lymphatic vessels. Sometimes tumor spreads through the lymphatics to surround airways (fig. 4-4), and can be picked up on a bronchial or transbronchial biopsy, causing confusion

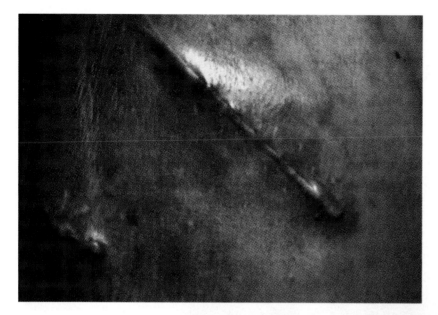

Figure 4-5

MALIGNANT MESOTHELIOMA

The tumor extends along a needle tract and also infiltrates a surgical scar. (Courtesy of Dr. D. Henderson, Adelaide, Australia.)

Figure 4-6

PERITONEAL MESOTHELIOMA

The diffuse tumor encases loops of bowel. (Fig. 4-13 from Fascicle 15, 3rd Series.)

with a primary lung cancer. A similar problem occurs when tumor grows into the mediastinum and permeates the lung through the hilum to surround the large airways.

Pleural mesotheliomas may grow into the chest wall, particularly along needle tracts or biopsy incisions, and then manifest as subcutaneous tumor nodules (fig. 4-5) (42). They can also grow through the mediastinum to enter the pericardium or surround mediastinal structures, and sometimes extend directly to the contralateral lung and pleura. Spread into and through the diaphragm, with superficial in-

volvement of underlying organs such as the liver, is not uncommon.

In the peritoneal cavity, tumor encasement of the bowel is seen in advanced disease (fig. 4-6), and tumor may spread into viscera, particularly into the wall of the bowel, into the omentum, and less frequently, into the retroperitoneum. In earlier disease, the tumor appears as multiple nodules rather than a rind (fig. 4-7). Peritoneal tumors can also grow along incisions into the subcutaneous tissue. On rare occasion, gross tumor is not visible in the peritoneal cavity and the patient presents with an apparent

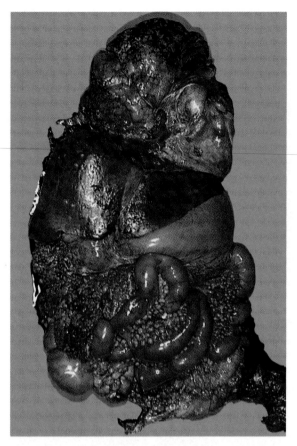

Figure 4-7

PERITONEAL MESOTHELIOMA

Tumor appears as multiple small nodules. Note the liver metastases. (Fig. 4-12 from Fascicle 15, 3rd Series.)

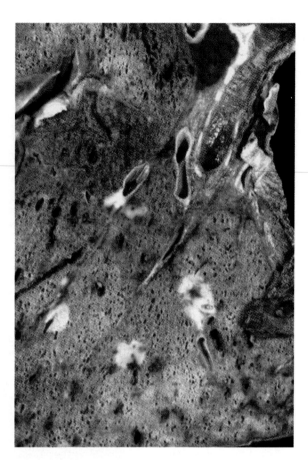

Figure 4-8

METASTASES FROM A PLEURAL MESOTHELIOMA

At autopsy, small tumor nodules are present in the contralateral lung.

inflammatory process, such as appendicitis, that shows foci of mesothelioma on microscopic examination (75). Peritoneal tumors may also grow through the diaphragm into the pleural cavity. With both pleural and peritoneal tumors, the bulk of the tumor is usually present in one body cavity or the other and the site of origin is obvious, but in some instances it is impossible to be sure where the tumor originated.

In the past, metastases were thought to be rare, but recent studies show that, at autopsy, metastases are found in at least 50 percent of patients with pleural mesothelioma (fig. 4-8) (20,42,142). They are rarely clinically apparent. The most common sites of metastases are hilar and mediastinal lymph nodes and the ipsilateral or contralateral lung. Widespread dissemination may occur, particularly with sarcomatous

lesions (20). Sarcomatous mesotheliomas have been reported to present with bone metastases (93; see Desmoplastic Mesotheliomas, below).

Like pleural mesotheliomas, peritoneal mesotheliomas may metastasize. Metastases are common at autopsy, but are not generally clinically important. Reported sites of metastases include abdominal, inguinal, and axillary lymph nodes, and a variety of visceral sites (71).

Primary pericardial mesotheliomas not only encase the heart but often invade the myocardium (fig. 4-9). They may spread into the pleura or mediastinum (170).

Mesotheliomas of the tunica vaginalis (fig. 4-10) are rare: a review in 1998 could find only 73 cases in the literature over the previous 30 years (134). Most of the cases were thought clinically to be hydroceles, often recurrent (68,132,134).

Figure 4-9

PERICARDIAL MESOTHELIOMA

The tumor has completely obliterated the pericardial space.

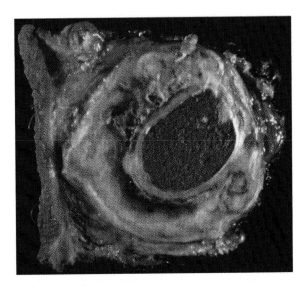

Figure 4-10

MALIGNANT MESOTHELIOMA
OF THE TUNICA VAGINALIS

The tumor completely surrounds the testis. (Courtesy of Dr. D. Grant, Surrey, BC, Canada.)

Gross findings include hemorrhagic fluid in the hydrocele sac, multiple papillary excrescences lining a hydrocele, or a hydrocele with prominent fibrous thickening of the wall. Some tumors grossly invade the spermatic cord or testis.

Microscopic Features

Because other tumors primary in or spreading to the serosal membranes may radiographically and grossly mimic a mesothelioma, microscopic examination is the only accurate method of diagnosis. Perhaps because so many malignant mesotheliomas in North America become medical-legal cases, immunohistochemical stains and other special techniques have assumed, in the minds of many pathologists, a disproportionate role in the diagnosis. But, as is true of all other types of tumor, the fundamental diagnosis of mesothelioma depends on routine hematoxylin and eosin (H&E) stains, and in most cases the diagnosis of mesothelioma is histologically obvious in routinely stained specimens. Immunohistochemical stains, histochemical stains, and electron microscopic evaluation are adjunctive techniques that must be applied to support or deny the impression gained from the H&E-stained images.

Despite the fact that these are highly aggressive tumors, epithelial mesotheliomas tend to

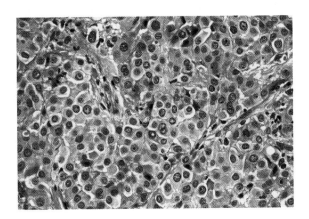

Figure 4-11

EPITHELIAL MALIGNANT MESOTHELIOMA:
BASIC APPEARANCE OF TUMOR CELLS

Epithelial mesotheliomas tend to be much more monotonous in appearance than carcinomas.

be relatively monotonous and often deceptively bland (fig. 4-11). This bland appearance can make separating benign from malignant mesothelial proliferations difficult (see chapter 5), but it does help in deciding whether a tumor is a mesothelioma or a carcinoma, since most carcinomas metastatic to the serosal membranes are much more pleomorphic. High-grade epithelial mesotheliomas occur but are distinctly uncommon. Sarcomatous forms of mesothelioma are more often cytologically atypical than epithelial forms, but many sarcomatous tumors appear remarkably innocuous. The number of mitoses is of no particular importance to the diagnosis since mitoses are typically scarce in epithelial mesotheliomas and can be prominent in reactive proliferations. Necrosis is generally not present, except in very large lesions or after radiation or chemotherapy. Squamous metaplasia is occasionally seen in epithelial mesotheliomas, but this phenomenon appears to have no special prognostic significance.

There are no standard grading schemes for mesothelioma. Indeed, for the most part it is not even clear how to grade "differentiation" in epithelial mesotheliomas, nor is it certain that the rare, cytologically high-grade epithelial mesothelioma (see below) is a more aggressive lesion. Because there exists the entity of *well-differentiated papillary mesothelioma* (see chapter 6), a tumor which is generally benign, we strongly advise against using the term "well differentiated" in reference to diffuse malignant mesothelioma since it only sows confusion.

CLASSIFICATION OF MORPHOLOGIC SUBTYPES OF MESOTHELIOMA

A variety of names exist in the literature for the different morphologic forms of mesothelioma. There is broad agreement that mesotheliomas can generally be divided into *epithelial, sarcomatous*, and *mixed epithelial and sarcomatous* (sometimes called *biphasic* or *bimorphic* [57]) forms.

There is some controversy regarding the precise terminology that should be used in the diagnosis of mesothelioma. Some authors prefer "epithelial" while others prefer "epithelioid." Similarly, some authors use "sarcomatous" while others prefer "sarcomatoid." According to Stedman's medical dictionary (161), "epithelium" refers to the layer covering all free surfaces, cutaneous, mucous, and serous. Thus the mesothelium is correctly referred to as an epithelium and tumors originating from the mesothelium with the appropriate histologic characteristics are properly referred to as "epithelial." Similarly, "sarcoma" according to Stedman's is a malig-

nancy of mesodermal tissues. Since mesothelium is derived from mesoderm (see chapter 1), tumors originating from the mesothelium with the appropriate histologic characteristics are properly referred to as "sarcomatous." The terminology used is somewhat a matter of preference. Although the recent World Health Organization Histological Typing of Lung and Pleural Tumors (171) uses the terms "epithelioid" and "sarcomatoid," for the reasons cited above, we prefer "epithelial" and "sarcomatous."

Within the sarcomatous group, most authors recognize *desmoplastic mesothelioma* as a distinctive category, but beyond this there is little consistency in the terms used to refer to the different epithelial and sarcomatous variants. There is also continuing confusion in terminology between entities that are clearly malignant mesotheliomas, for example, sarcomatous mesothelioma, and older named entities such as *fibrous mesothelioma*, which refers both to sarcomatous mesothelioma and to solitary fibrous tumor (see chapter 6). We strongly discourage the use of the term fibrous mesothelioma because of this problem.

Epithelial forms predominate in almost all studies. In a summarized series of 382 pleural tumors, 55 percent were epithelial, 24 percent mixed, and 22 percent sarcomatous (86). Similar proportions are seen in more recent reports (57,144). In the peritoneal cavity, an even higher proportion of epithelial forms are found. Kannerstein and Churg (70) examined 82 cases and classified 75 percent as epithelial, 24 percent as mixed, and only 1 percent as sarcomatous. Our own experience and that of others (12) also show that purely sarcomatous tumors are rare in the peritoneal cavity. Separation into even the three general categories is somewhat arbitrary, however, because in some instances it is difficult to decide whether a spindle cell component is part of the stroma or part of the tumor, and even whether individual cells should be classified as epithelial or sarcomatous (see below). As well, there are no generally accepted rules for what percentage of the tumor must be of an epithelial or sarcomatous pattern before classifying a tumor as mixed. We recommend that mesotheliomas be designated as mixed if there are clearly both sarcomatous and epithelial elements, taking care

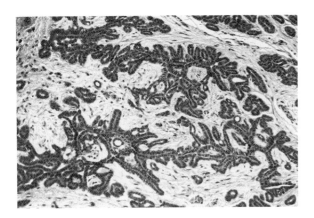

Figure 4-12

EPITHELIAL MALIGNANT MESOTHELIOMA

Glands are lined by low cuboidal cells and form outward branches, part of the spectrum of the tubulopapillary pattern of epithelial mesothelioma.

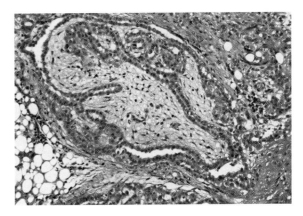

Figure 4-13

EPITHELIAL MALIGNANT MESOTHELIOMA

The tubulopapillary pattern demonstrates both outward branches and complex invaginations.

to distinguish reactive stroma from a true sarcomatous component.

The importance of classification obviously lies in the prognostic information it produces. In this regard, the separation of epithelial and sarcomatous forms is important because there are distinct differences in survival for patients with these two types of tumor (see Treatment and Prognosis) and experience has shown that sarcomatous tumors are not amenable to radical triple modality therapy (164,192).

Subclassifying the epithelial forms does not appear to convey any particular prognostic or behavioral information. To complicate matters, mesotheliomas are typically quite polymorphous, so that a given tumor may show two or more different patterns, either within a type (e.g., all epithelial) or between types (e.g., mixed epithelial and sarcomatous). In fact, epithelial mesotheliomas with only one histologic pattern are the exception rather than the rule. We have, therefore, written our description of the epithelial form to indicate particular names that have been associated with each pattern, but we do not intend to make this a formal classification. The names at this point are simply convenient mental markers to remind the pathologist that mesotheliomas can have a particular appearance.

Epithelial Malignant Mesothelioma

Basic Cell Morphology. Epithelial mesotheliomas are, in general, remarkable for their de-

ceptive blandness. The cells of most epithelial mesotheliomas are cuboidal or polygonal to flattened and show little variation from one to the next (fig. 4-11; see also chapter 5 for further illustrations). The nuclei also tend to be monotonous, although usually many or all have large nucleoli (fig. 4-11), but in some cases prominent nucleoli are not obvious (see chapter 5). Nucleoli may be difficult to detect in flattened cells and such tumors require careful examination. In many forms of epithelial mesothelioma the cell borders are sharply demarcated, and when these form sheets there is a pavement-like appearance (fig. 4-11). Glands lined by flattened cells with outwardly branching tubules (tubulopapillary pattern) are typical of epithelial mesotheliomas (figs. 4-12, 4-13). Papillary structures with fibrovascular cores or tumor cell cores are common. In most epithelial mesotheliomas, mitoses are hard to find.

In contrast, the cells of metastatic carcinoma, the major differential diagnostic consideration, typically are much more pleomorphic. Carcinoma cells are generally cuboidal or polygonal to columnar, and the presence of only columnar cells is a finding against a diagnosis of mesothelioma. The nuclei of metastatic carcinoma cells are generally fairly pleomorphic and mitoses may be numerous, again not features of epithelial mesothelioma. Glands with very flattened cells, a common phenomenon in mesothelioma, are uncommon in carcinomas. In most carcinomas,

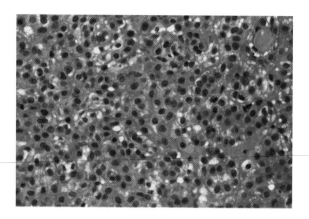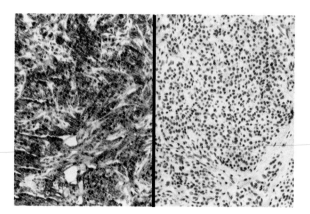

Figure 4-14

EPITHELIAL MALIGNANT MESOTHELIOMA

Left: The tumor is composed of polygonal cells, some with cytoplasmic vacuoles.

Right: Staining is strongly positive for calretinin (left) and negative for carcinoembryonic antigen (CEA) (right), a typical result in malignant mesothelioma.

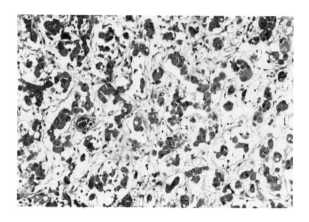

Figure 4-15

EPITHELIAL MALIGNANT MESOTHELIOMA

This tumor is composed of polygonal cells, some with cytoplasmic vacuoles, infiltrating a myxoid stroma.

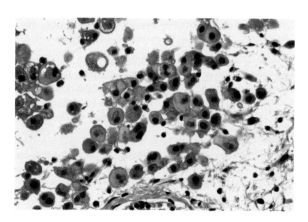

Figure 4-16

EPITHELIAL MALIGNANT MESOTHELIOMA

The lesion is composed of very large rounded cells resembling the reactive mesothelial cells seen in effusions. The tumor infiltrates a myxoid stroma containing hyaluronate, which is secreted by the tumor cells.

gland proliferation leads to a gland within gland pattern rather than outwardly branching tubules, although exceptions do occur. These are not absolute rules by any means and malignancies metastatic to the serosal membranes can certainly mimic mesothelioma, but in most instances the fundamental cell types are quite different.

Morphologic Forms. Many epithelial mesotheliomas have sheets of large polygonal cells in the pattern described above (figs. 4-11, 4-14). This pattern is sometimes referred to as *epithelioid* (12,30,57). These tumors may infiltrate fi-

brous tissue, fat, and muscle in sheets, or on a cell by cell basis (figs. 4-15, 4-16), and can mimic the reactive mesothelial cells seen in cytologic effusions, although a direct comparison shows that the tumor cells are much larger. Benign mesothelial cells, of course, do not infiltrate fat and muscle (see chapter 5).

A closely related morphologic variant has been labeled *deciduoid mesothelioma* (154,168). This pattern consists of large polygonal cells with abundant eosinophilic cytoplasm (fig. 4-17) that somewhat resemble decidua. Initially,

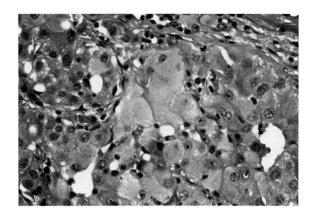

Figure 4-17

EPITHELIAL MALIGNANT MESOTHELIOMA

This tumor shows a so-called deciduoid pattern, with very large polygonal cells that somewhat resemble decidua.

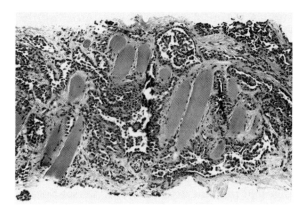

Figure 4-18

EPITHELIAL MALIGNANT MESOTHELIOMA

This tumor has a partially tubulopapillary pattern and infiltrates the muscle of the chest wall.

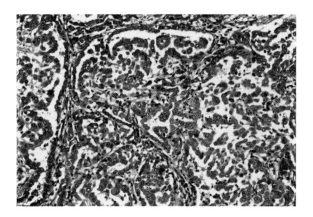

Figure 4-19

EPITHELIAL MALIGNANT MESOTHELIOMA

This area of the tumor shown in figure 4-15 shows another variant of the tubulopapillary pattern. Multiple patterns are a common finding in malignant mesothelioma.

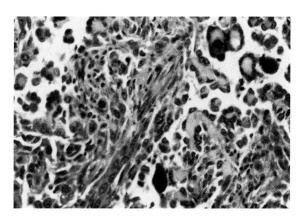

Figure 4-20

EPITHELIAL MALIGNANT MESOTHELIOMA

Psammoma bodies are present.

these lesions were found in the peritoneal cavity of women and were thought to be a bizarre benign decidual reaction (168). Immunohistochemical and ultrastructural examination showed that the cells were mesothelial and the tumors behaved in a malignant fashion. Further experience has shown that deciduoid mesotheliomas also arise in the pleura and in men, and that they behave like other epithelial mesotheliomas (154). Tumors comprised only of this pattern are rare, but it is not unusual to see a transition from the typical polygonal epithelial form to a focally deciduoid form.

The other common pattern of epithelial mesothelioma is often referred to as *tubulopapillary*. In this variant papillae are covered by a single layer of cuboidal to flattened cells, and glands are lined by similar cells, often with outwardly branching tubules (figs. 4-12, 4-13, 4-18, 4-19). Some tumors consist only of papillae, others only of glands, but usually a combination of the two is observed. Psammoma bodies may be seen in tumors with this histologic appearance, but are infrequent (fig. 4-20). There appears to be a tendency to regard mesotheliomas with a tubulopapillary pattern as "well differentiated" (57), but it is not clear whether survival data support such a claim. Tumors with a pure tubulo-papillary pattern are not common; usually there is a mixture of the tubulopapillary and other patterns (figs. 4-15, 4-19).

The cells of mesotheliomas with a tubulo-papillary or epithelioid pattern, particularly the epithelioid form, may contain cytoplasmic droplets that resemble the mucin droplets of carcinomas (fig. 4-21), sometimes even mimicking signet ring cell carcinoma. The droplets are negative with stains for neutral mucin such as digested periodic acid–Schiff (dPAS; see Special

Techniques, below), and this finding is helpful in confirming a diagnosis of mesothelioma.

Epithelial mesotheliomas may also form a *microcystic* (also called *adenomatoid*) pattern, which usually is composed of flattened or attenuated cells lining glands that contain pale hematoxyphilic material called hyaluronate (fig. 4-22, left; see Special Techniques). This material stains with alcian blue, pH 2.5, and can be removed by pretreatment with hyaluronidase (fig. 4-22, right). This pattern of staining is a useful confirmatory test for the diagnosis of mesothelioma. Microcystic tumors can form very bland, lace-like patterns and may be mistaken for adenomatoid tumors; however, careful examination usually shows the presence of large nucleoli, and the tumor cells are distinctly infiltrative as well as being diffuse. Only a handful of adenomatoid tumors have been reported in the pleura and these have all been small localized nodules (72, and see chapter 6). As a rule, lesions in the pleura with the appearance described above are malignant mesotheliomas. Occasionally, lesions with the microcystic pattern resemble adenoid cystic carcinomas of the salivary glands, and in other instances, large lakes of hyaluronate are formed. On gross examination, these tumors show cysts containing mucoid material. A stroma of hyaluronate is sometimes seen in epithelial mesotheliomas and tumor cells may appear to float in the stroma (see figs. 4-15, 4-16), similar to

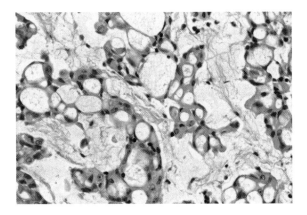

Figure 4-21

EPITHELIAL MALIGNANT MESOTHELIOMA

The tumor is composed of markedly vacuolated cells in a pattern somewhat resembling a benign adenomatoid tumor. The vacuoles did not stain for neutral mucin, a helpful feature in separating this tumor from a carcinoma.

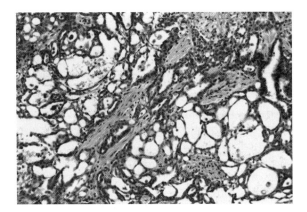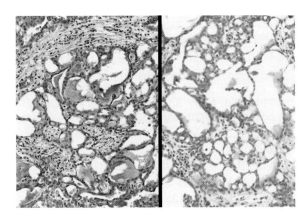

Figure 4-22

EPITHELIAL MALIGNANT MESOTHELIOMA

Left: This tumor has a microcystic pattern.

Right: The tumor is stained with alcian blue, pH2.5, without (left) or with (right) prior hyaluronidase treatment. The production of hyaluronate is characteristic of many epithelial mesotheliomas.

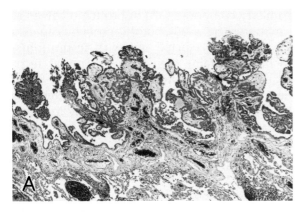

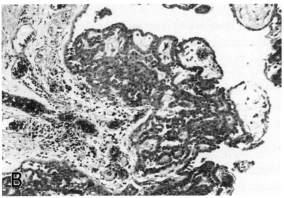

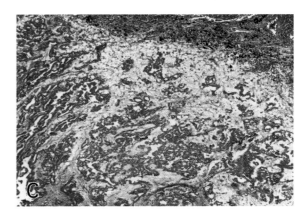

Figure 4-23

EPITHELIAL MALIGNANT MESOTHELIOMA

A: Example of complex surface tumor.

B: Higher-power view of the surface proliferation. The complexity of the process raises the suspicion of mesothelioma. But without invasion, mesothelioma should not be diagnosed.

C: Another area of this tumor demonstrates typical invasive malignant mesothelioma, thus confirming the diagnosis.

mucinous carcinoma of the breast or colon. The stroma, however, does not stain for neutral mucin (as opposed to mucinous carcinoma) but does stain with low pH alcian blue or colloidal iron (see Special Techniques).

A small number of cases labeled *small cell mesothelioma* have been reported (95). These tumors are composed, in part, of cells that are fairly small and superficially resemble cells of small cell carcinoma. The mesothelial cells are arranged in monotonous sheets and do not form the usual ribbons and rosettes seen in small cell carcinoma. They are keratin positive, and also usually stain for neuron-specific enolase, but not chromogranin. The mitotic index is low, and the chromatin open, as opposed to small cell carcinomas. Crush artefact and basophilic staining of blood vessels, characteristic of small cell carcinoma, are similarly absent. Most important, other areas of the tumor always show more typical patterns of epithelial mesothelioma if multiple sections are examined.

Epithelial malignant mesotheliomas, especially the papillary form, often grow along the free serosal surface. Because the distinction be-

tween reactive mesothelial proliferations and mesotheliomas growing on the free surface is problematic, we do not recommend making a diagnosis of malignant mesothelioma unless clearly invasive tumor is present (fig. 4-23). This problem is discussed at length in chapter 5. To further complicate the diagnosis, papillary surface proliferations with broad fibrovascular cores can, in individual fields, mimic well-differentiated papillary mesothelioma, a generally benign tumor (see chapter 6); however, the presence of invasive tumor makes the diagnosis of malignant mesothelioma obvious (fig. 4-23).

Epithelial mesotheliomas are often associated with a marked, dense, *desmoplastic* response, especially in the pleura. Unfortunately, this is not diagnostically useful, since metastatic carcinomas may evoke a similar response (see Differential Diagnosis), and dense fibrosis is also fairly common in benign reactive mesothelial proliferations. Occasionally, the desmoplastic response traps the cells of epithelial mesothelioma in a way that mimics metastatic carcinoma, particularly metastatic breast carcinoma (fig. 4-24).

Epithelial mesotheliomas sometimes are composed of *clear cells* (fig. 4-25). In most instances, areas of more ordinary appearing mesothelioma are present as well, but on occasion, such tumors show only clear cells and need to be distinguished from metastatic renal cell carcinoma.

The microscopic appearance of metastatic mesothelioma is usually similar to that of the parent tumor, although numerous morphologic variants may be present in any given patient. Sometimes metastatic mesothelioma in the lung closely mimics a carcinoma with a micro-

papillary component (3) or bronchioloalveolar carcinoma (fig. 4-26), but this problem can usually be readily sorted out with immunohistochemical stains (fig. 4-26; see Special Techniques, below). As noted previously, mesotheliomas can grow into the lung along interlobular septa or hilar structures and may infiltrate the airway wall, causing confusion with a primary bronchogenic carcinoma (fig. 4-27).

Mucin-Positive Mesothelioma

As described in the section on Special Techniques, mesotheliomas do not ordinarily produce dPAS-positive neutral mucins, the typical finding in adenocarcinomas. Alcian blue, pH 2.5, and colloidal iron sometimes stain hyaluronate droplets in mesotheliomas, but as a rule this staining is removable by hyaluronidase pretreatment (fig. 4-22) (54). Hammar et al. (54), however, reported three examples of epithelial mesothelioma in which the tumor cells produced mucicarmine-, dPAS-, and alcian blue–positive droplets that were resistant to hyaluronidase (fig. 4-28A,B). The tumors were otherwise histologically typical mesotheliomas. Ultrastructurally, they showed distinctive crystalloids in the droplets or in the extracellular space (fig. 4-28C,D), and this material, which the authors suggested was crystalized proteoglycan, appeared to be responsible for the positive staining results.

Figure 4-24

EPITHELIAL MALIGNANT MESOTHELIOMA

This tumor shows a desmoplastic stromal response. Note the potential for confusion with metastatic carcinoma, in particular with a lobular carcinoma of breast.

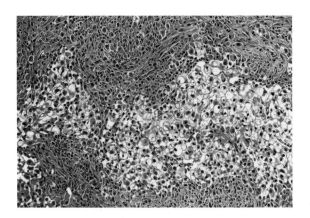
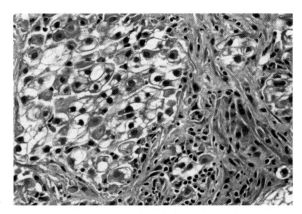

Figure 4-25

EPITHELIAL MALIGNANT MESOTHELIOMA

Left: The clear cells of this tumor are seen.

Right: Higher-power view. Occasionally, epithelial mesotheliomas are composed entirely of clear cells, adding metastatic renal cell carcinoma to the differential diagnosis.

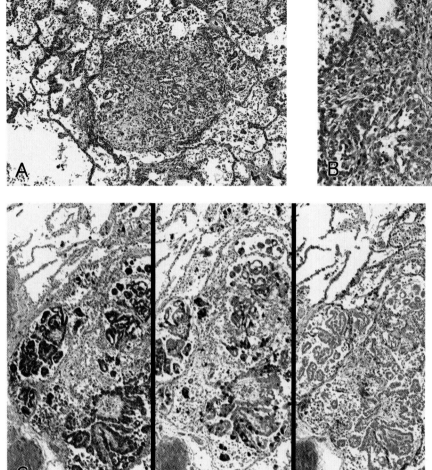

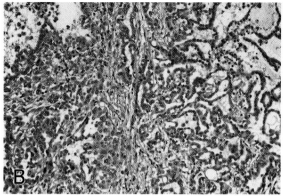

Figure 4-26

EPITHELIAL MALIGNANT MESOTHELIOMA

A: Metastatic tumor in lung mimics a bronchioloalveolar carcinoma.

B: Another focus of metastasis of the tumor shown in figure A.

C: The immunohistochemical staining pattern confirms the diagnosis of mesothelioma. There is strong staining for calretinin (left) and cytokeratin (CK) 5/6 (middle). CEA (right) is negative in the tumor cells and positive in the normal alveolar epithelial cells (left, center, right).

Figure 4-27

EPITHELIAL MALIGNANT MESOTHELIOMA

The tumor infiltrates a large airway. Mesotheliomas gain access to hilar structures, including airway walls, by invading from the surface and on a small biopsy can be confused with carcinoma.

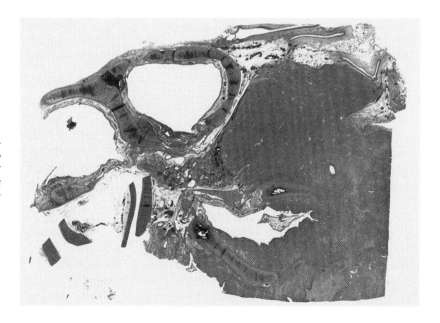

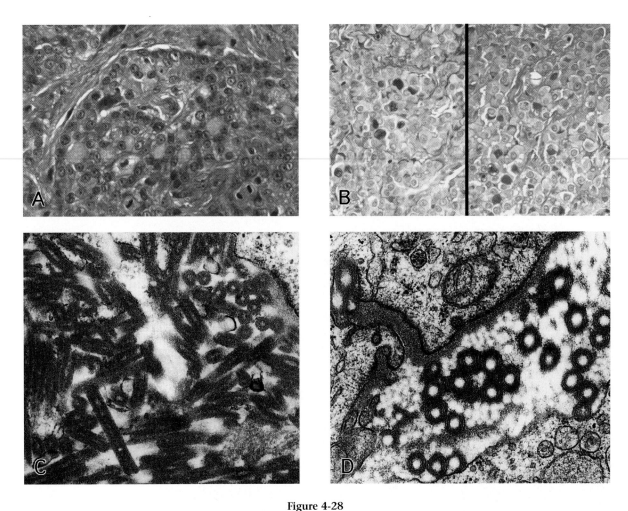

Figure 4-28

EPITHELIAL MALIGNANT MESOTHELIOMA: MUCIN-POSITIVE VARIANT

A: The tumor looks like an ordinary epithelial mesothelioma but has pale-staining cytoplasmic vacuoles. (Courtesy of Dr. S. Hammar, Bremerton, WA.)

B: Alcian blue, pH 2.5, stains the cytoplasmic vacuoles (left). The staining did not disappear with hyaluronidase treatment (right).

C: Electron micrograph shows crystalline structures of mucin-positive material in a longitudinal section. This is a characteristic finding in mucin-positive mesothelioma.

D: Electron micrograph shows that the crystalline material has a scroll-like pattern on cross section.

High-Grade (Pleomorphic) Epithelial Mesothelioma

Most epithelial malignant mesotheliomas are cytologically monotonous and remarkably bland in their appearance. There are rare epithelial mesotheliomas, however, that are cytologically of high grade and have considerable cell to cell variation (figs. 4-29, 4-30). Such tumors are difficult to distinguish from metastatic carcinomas. Immunohistochemistry and electron microscopy are not always helpful; the tumors may not stain for any antigen except broad spectrum keratin, and they may have no specific ultrastructural features. In our experience, most such high-grade lesions are in fact metastatic carcinomas, but in a given case, the issue may not be resolvable based on a small biopsy specimen.

Sarcomatous Mesothelioma

Like epithelial mesotheliomas, sarcomatous tumors can have a wide range of appearances, and combinations of different patterns may be found within the same tumor. The spindle cells of sarcomatous mesothelioma may be very plump,

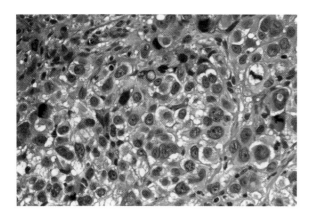 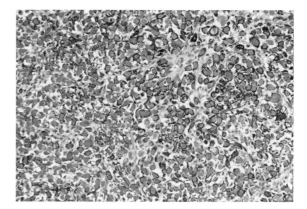

Figure 4-29

HIGH-GRADE EPITHELIAL MESOTHELIOMA

Left: There are numerous mitoses and a large degree of nuclear pleomorphism. The diagnosis of high-grade epithelial mesothelioma requires confirmatory immunochemical staining.

Right: Strong diffuse staining for CK5/6. Tumor cells were also positive for calretinin and negative for CEA and LeuM1.

sometimes verging on an epithelial configuration (fig. 4-31), or relatively long and thin with sparse cytoplasm (fig. 4-32). Cytologically, sarcomatous mesotheliomas can be quite bland, but, as opposed to epithelial mesotheliomas, it is not unusual to find high-grade cytology and numerous mitoses (fig. 4-32).

The most basic form of sarcomatous mesothelioma is a lesion composed of closely packed spindled cells that may vary from very bland to moderately anaplastic. The lesions often form a storiform pattern and resemble malignant fibrous histiocytoma (fig. 4-32). Tumor giant cells are sometimes seen (fig. 4-32). Other examples resemble fibrosarcomas, with interlacing bundles of cells. All of these forms are very cellular and look like sarcomas in other sites. Heterologous elements, usually malignant cartilage and malignant bone, are found in a small proportion of tumors (fig. 4-33). Such tumors may appear calcified on CT scan. Histologically, the heterologous foci may be indistinguishable from osteosarcoma or chondrosarcoma (fig. 4-33), and, rarely, leiomyosarcoma or rhabdomyosarcoma.

Most ordinary sarcomatous mesotheliomas are broad-spectrum keratin positive, although the staining may be focal. This positivity is very helpful in supporting the diagnosis (figs. 4-33, right; 4-34); however, there are occasional keratin-negative tumors (see Special Techniques, below). Calretinin and cytokeratin (CK) 5/6 may also be positive in sarcomatous mesothelioma,

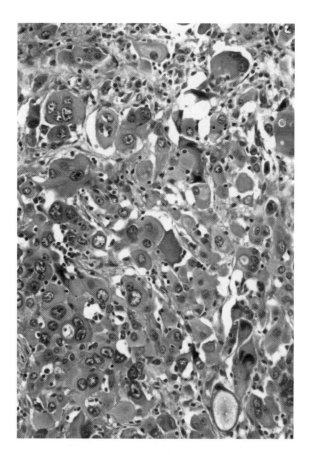

Figure 4-30

HIGH-GRADE EPITHELIAL MESOTHELIOMA

Giant cells and marked pleomorphism are present. This tumor was calretinin and CK5/6 positive and negative for CEA and thyroid transcription factor (TTF)1.

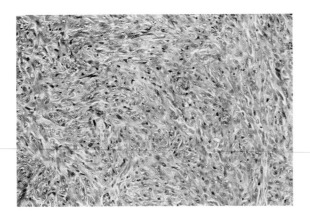
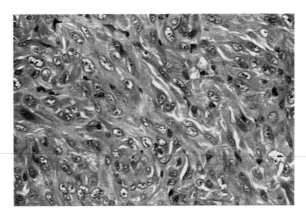

Figure 4-31

SARCOMATOUS MESOTHELIOMA

Left: The tumor is composed of plump spindled cells with relatively little atypia.
Right: High-power view.

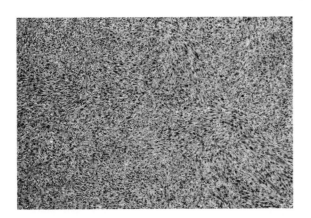
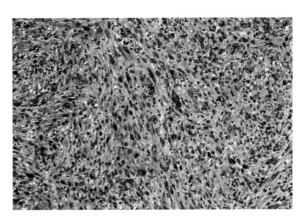

Figure 4-32

SARCOMATOUS MESOTHELIOMA

Left: Low-power view shows a storiform pattern that resembles malignant fibrous histiocytoma.
Right: High-power microscopy shows the high-grade nuclei and occasional giant cells, fairly common findings in sarcomatous mesothelioma in contrast to epithelial mesothelioma.

although much less frequently than keratin positivity. Tumors with osteosarcomatous and chondrosarcomatous foci may also be keratin positive (fig. 4-33, right) (190), and positivity may be present in the heterologous elements.

In general, all of these lesions are simply referred to as sarcomatous mesothelioma, without further specific names applied. As noted above, they should not be called "fibrous mesothelioma" because the same term has been used repeatedly in the past for solitary fibrous tumor, and instances of confusion between a malignant and a benign lesion are still common when this wording is employed.

There are three variant forms of sarcomatous mesothelioma that have specialized morphologic features. *Transitional mesothelioma* (12,53, 91) is a term applied to tumors in which the cells are spindled and very plump, so that, depending on the plane of section, it is difficult to decide whether the process is epithelial or sarcomatous (fig. 4-35). In our experience, these tumors usually have an astonishingly bland cytologic appearance. Some authors have labeled these tumors "poorly differentiated," however; emphasizing that "differentiation" as a predictor of behavior is not a concept that is easily applied to mesothelioma. In our experience, the

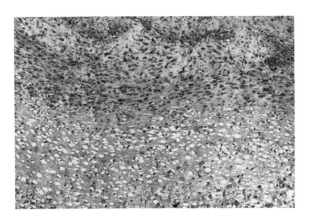
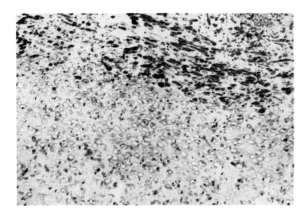

Figure 4-33

SARCOMATOUS MESOTHELIOMA

Left: A focus of cartilaginous differentiation is seen.
Right: Staining for broad-spectrum keratin is seen in the spindle and cartilage cells.

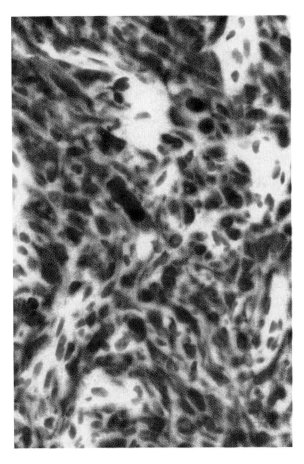

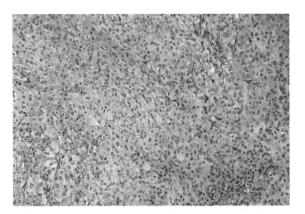

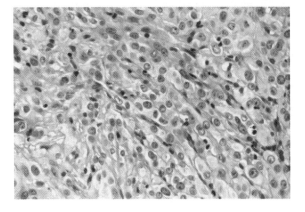

Figure 4-35

**TRANSITIONAL VARIANT OF
SARCOMATOUS MESOTHELIOMA**

Top: The tumor cells are not clearly epithelial or spindled in form. Despite the bland appearance, necrosis is present (lower right).
Bottom: The appearance is bland.

Figure 4-34

SARCOMATOUS MESOTHELIOMA

Diffuse staining for broad-spectrum cytokeratin.

Figure 4-36

LYMPHOHISTIOCYTOID MALIGNANT MESOTHELIOMA

A: The tumor resembles a large cell lymphoma.
B: Higher-power view.
C: Broad-spectrum keratin stain.

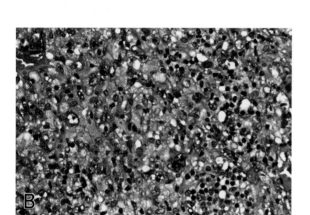

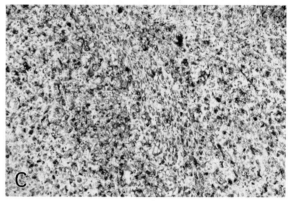

usual bland-appearing transitional mesothelioma behaves like the other forms of sarcomatous mesothelioma.

Lymphohistiocytoid malignant mesothelioma was first described by Henderson et al. (56) and additional cases have been subsequently reported (77). These tumors have an intense inflammatory background of small lymphocytes, plasma cells, and sometimes eosinophils. Mixed into the inflammatory component are large polygonal to slightly spindled tumor cells containing nuclei with vesicular chromatin and large nucleoli (fig. 4-36). The picture strongly suggests a malignant lymphoma. The tumor cells, however, are positive for keratin (fig. 4-36C) and sometimes for epithelial membrane antigen, calretinin, or CK 5/6, and do not stain for lymphoid markers. Ultrastructural examination occasionally shows evidence of epithelial mesothelial differentiation (57). Although the initial reports (56,57) suggested that lymphohistiocytoid mesothelioma behaves in the aggressive fashion of sarcomatous mesothelioma, recent data from the French Mesopath Group show that survival in patients with these tumors is identical to that

of those with epithelial mesothelioma (F. Galateau-Salle, personal communication).

Desmoplastic malignant mesothelioma (DMM) was first described by Kannerstein and Churg (71) and additional cases have been subsequently published (20,93,184). DMM is a variant of sarcomatous mesothelioma in which most of the tumor, by definition, consists of paucicellular, densely collagenized tissue. These tumors almost always occur in the pleural cavity. Grossly, they look like other forms of mesothelioma. Microscopically, the paucicellular parts usually show a storiform, or "patternless pattern" of Stout, an appearance with elongated, often very attenuated, tumor cells present in the interstices between the collagen bundles (fig. 4-37) (162a). Often these lesions occupy a vastly thickened pleura and at first glance appear to be benign fibrous tissue (fig. 4-37, left), and the differential diagnosis is, in fact, organizing pleuritis or other benign pleural lesions such as plaques (see chapter 5).

Unfortunately, the storiform and paucicellular patterns, although suggestive of DMM, are also seen in benign reactions such as organizing pleuritis (see chapter 5). For that reason, a

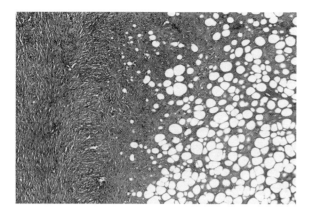 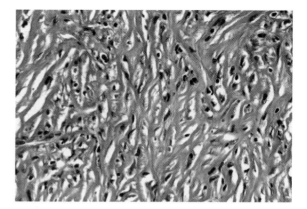

Figure 4-37

DESMOPLASTIC MESOTHELIOMA

Left: There is a paucicellular, highly collagenized pattern and extensive fat infiltration. Fat in the upper right portion of the field has undergone bland necrosis (see figure 4-42).

Right: High-power view shows the "patternless pattern," one of the common appearances of desmoplastic mesothelioma. Nuclear detail is difficult to discern, as is typical of many desmoplastic mesotheliomas.

reliable diagnosis of DMM requires the proper paucicellular pattern plus one of the following: stromal invasion, bland necrosis, overtly sarcomatous foci, or distant metastases (31,93). Stromal invasion is usually represented by infiltration of the fat of the chest wall by spindle cells, but tumor cells can also infiltrate muscle, bone, and lung (figs. 4-38–4-41). The tumor cells insinuate themselves in a characteristic fashion, isolating individual fat cells and often forming microscopic tumor cell spindles between them (fig. 4-38). DMM cells infiltrating fat are, for all practical purposes, always broad-spectrum keratin positive (fig. 4-38C). Benign cells in any mesothelial reaction can also be keratin positive, but keratin-positive benign spindle cells usually do not invade adipose tissue (see chapter 5). Invasion of DMM along interlobular septa (fig. 4-40) is common and can produce a very confusing diagnostic picture. When DMM invades the lung, the spindle cells proliferate in the airspaces, sometimes producing a pattern that mimics organizing pneumonia (fig. 4-41) (31,93).

Bland necrosis is necrosis without any cellular reaction and usually little or no karrhyorhexis. At low-power magnification, this is best detected as sharply demarcated areas with a slightly different pink staining pattern in a typical paucicellular region. It can also occur in the chest wall fat (figs. 4-37, left; 4-42).

Overtly sarcomatous foci occur when the cellularity increases to the point that, focally, the lesion is clearly a sarcoma that is indistinguishable from ordinary sarcomatous mesothelioma. These foci may be small or widely scattered in a given specimen (fig. 4-43). An area of densely packed spindle cells deep to the pleural surface in an apparently fibrotic pleural reaction should raise the suspicion of DMM.

Occasionally, DMM presents with metastatic disease that histologically may be quite similar to the primary (fig. 4-44). These distant metastases are keratin positive. Metastases to the lung also may be seen (figs. 4-10, 4-45).

An additional feature, which has not been well described in the literature, is the presence of expansile stromal nodules (fig. 4-46). As noted above, they can be seen in epithelial mesotheliomas, but they also occasionally are found in DMM. They have pushing borders and, usually, a different tinctorial character compared to the surrounding stroma. Expansile nodules are not a feature of organizing pleuritis. Heterologous elements, usually osteoid, may be seen in DMM, but this is rare (fig. 4-47).

It may be extremely difficult to diagnose DMM and characteristic histologic foci are sometimes small and widely scattered. For this reason we suggest that when there is a clinical suspicion of mesothelioma, an apparently fibrous process should be widely sampled histologically.

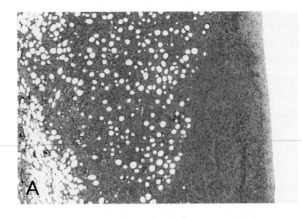

Figure 4-38

DESMOPLASTIC MESOTHELIOMA

A: Low-power view shows extensive infiltration of the fat of the chest wall.

B: Higher-power view shows spindled cells growing between fat cells.

C: Broad-spectrum cytokeratin–positive cells infiltrate fat.

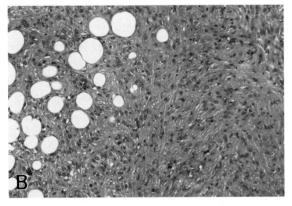

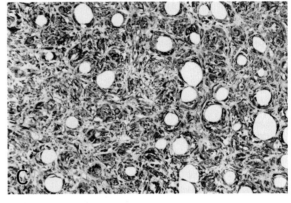

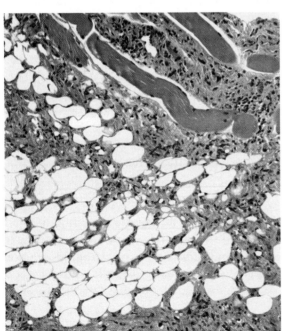

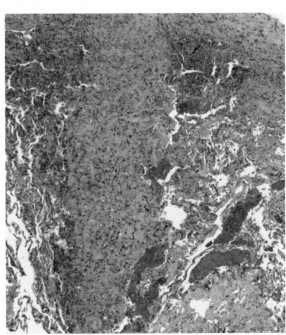

Figure 4-39

DESMOPLASTIC MESOTHELIOMA

Tumor cells infiltrate fat and skeletal muscle of the chest wall.

Figure 4-40

DESMOPLASTIC MESOTHELIOMA

The tumor grows along the pleural surface and down an interlobular septum.

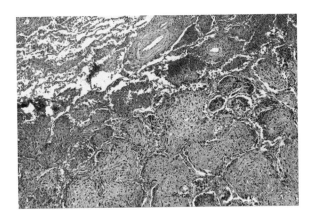

Figure 4-41

DESMOPLASTIC MESOTHELIOMA

The tumor invades the lung and mimics organizing granulation tissue.

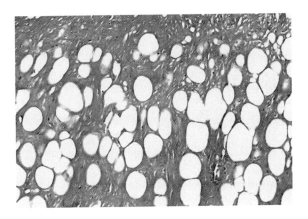

Figure 4-42

DESMOPLASTIC MESOTHELIOMA

Bland necrosis in another area of the tumor shown in figure 4-37. This type of necrosis can involve both solid areas of tumor and tumor infiltrating fat.

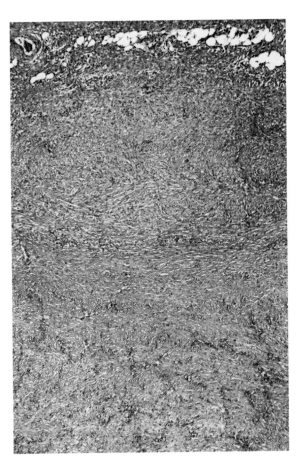

Figure 4-43

DESMOPLASTIC MESOTHELIOMA

Left: Low-power view shows a paucicellular storiform pattern. Most of the tumor had this appearance.
Right: A sarcomatous focus is seen.

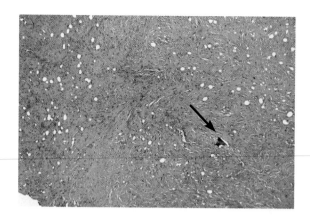
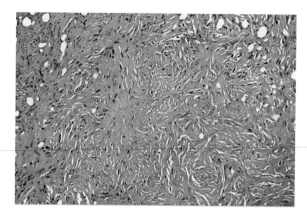

Figure 4-44

METASTATIC DESMOPLASTIC MESOTHELIOMA

Left: Desmoplastic mesothelioma presents as a pathologic fracture of the femur. Note the bone fragment (arrow).
Right: Higher-power view shows a pattern that is histologically identical to that seen in desmoplastic mesothelioma in the pleura. This lesion was positive for broad-spectrum keratin.

Figure 4-45

METASTATIC DESMOPLASTIC MESOTHELIOMA

A nodule of metastatic tumor is seen in the lung.

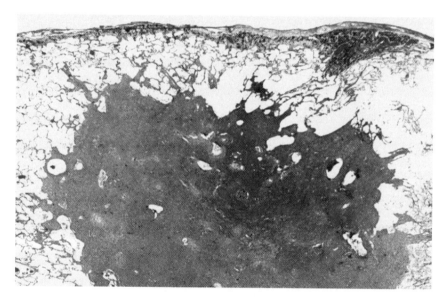

Mixed Epithelial and Sarcomatous Mesothelioma

Mixed epithelial and sarcomatous tumors are a combination of an epithelial and a sarcomatous component. Any combination, including mixtures of epithelial and desmoplastic patterns (184), is possible (figs. 4-48–4-50). Mixed tumors often have more than one epithelial pattern. The transition from the epithelial component to the spindle cell component can be abrupt or extremely gradual. Some tumors are composed of large areas of one component and separate large areas of another component (100,101), whereas other tumors show an intimate admixture of the two.

Mesothelioma of the Tunica Vaginalis

About two thirds of the reported mesotheliomas of the tunica vaginalis are epithelial, and almost all of the remaining are mixed epithelial and sarcomatous (68,132,134). Histologically, they are identical to pleural and peritoneal lesions. Mesotheliomas of the tunica vaginalis may spread locally, invading the testis and epididymis, and can metastasize to inguinal lymph nodes or more distantly. They may also extend

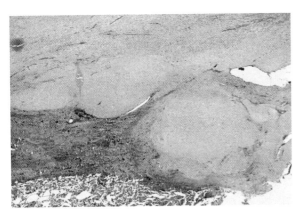 

Figure 4-46

DESMOPLASTIC MESOTHELIOMA

Left: The expansile stromal nodules seen here are not seen in benign reactive processes.
Right: Higher-power view of an expansile nodule. Note the deceptively low cellularity.

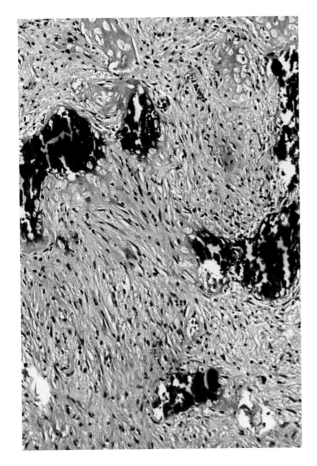 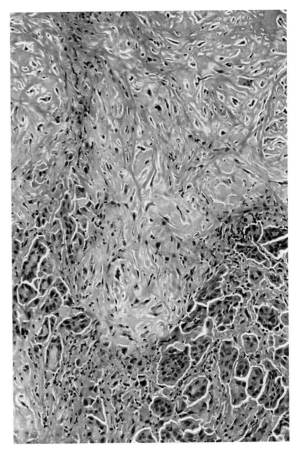

Figure 4-47

DESMOPLASTIC MESOTHELIOMA

There is formation of osteoid.

Figure 4-48

**MIXED EPITHELIAL AND
SARCOMATOUS MESOTHELIOMA**

The sarcomatous component has the features of a desmoplastic mesothelioma.

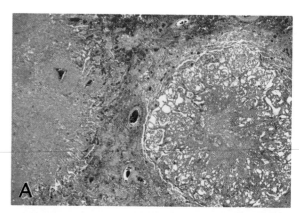

Figure 4-49

**MIXED EPITHELIAL AND
SARCOMATOUS MESOTHELIOMA**

A: Low-power view shows a combination of microcystic (adenomatoid) and desmoplastic patterns.
B: High-power view of the microcystic component.
C: High-power view of the desmoplastic component.

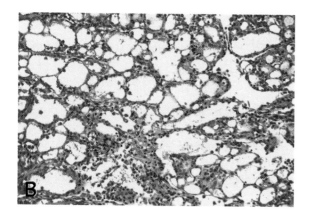

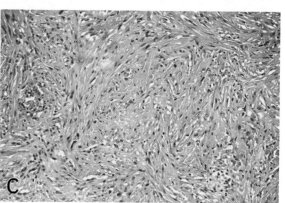

Figure 4-50

**MIXED EPITHELIAL
AND SARCOMATOUS
MESOTHELIOMA**

The sarcomatous component has osteosarcomatous-like foci.

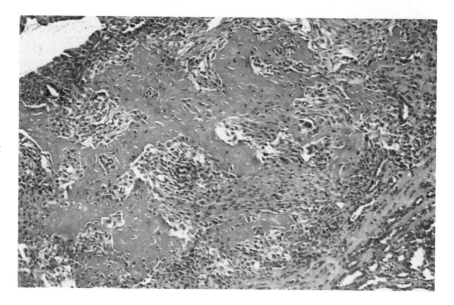

to involve the peritoneal cavity more diffusely. Separation of mesothelioma of the tunica vaginalis from reactive mesothelial proliferations in hydroceles is sometimes problematic; some guidelines for making this distinction are presented in chapter 5.

Pericardial Mesothelioma

Epithelial, mixed, and sarcomatous forms of pericardial mesothelioma have been reported (170). Histologically, they are similar to mesotheliomas in other sites.

SPECIAL TECHNIQUES FOR DIAGNOSIS

Mesothelioma may be difficult to distinguish from other malignancies involving the serosal membranes, especially in small biopsy specimens. Because of this many special techniques are used as aids to diagnosis.

Mucin Stains

Mucin stains are very useful for diagnosing mesothelioma if they are performed and interpreted correctly. The stain of choice to demonstrate neutral mucin is dPAS, the hallmark of carcinoma, and neutral mucin should not be found in mesotheliomas (there are rare mucin-positive mesotheliomas that have some special features; see Mucin-Positive Mesothelioma, above). The substance in question must stain as mucin droplets and not lysosomes or some other nonspecific feature.

Mucicarmine stains are difficult to interpret in this setting because mesotheliomas often make hyaluronate and this sometimes stains with mucicarmine. We recommend avoiding mucicarmine for this reason, but if it is used and "mucin droplets" are found in a tumor that appears to be a mesothelioma, the stain should be repeated with hyaluronidase pretreatment. In almost all instances, the staining reaction will disappear in mesotheliomas.

Alcian blue, pH2.5, or colloidal iron stains with and without hyaluronidase can demonstrate hyaluronate production in a mesothelioma, but in many tumors the hyaluronate washes out in the processing, so false negatives are frequent. With rare exceptions (see Mucin-Positive Mesothelioma), the alcian blue–positive mucin-like material in mesotheliomas disappears with hyaluronidase pretreatment; its persistence indicates that the tumor is a carcinoma. Positive alcian blue stains may appear as mucin droplets in tumor cells, as lakes of material in the spaces of microcystic forms of mesothelioma (see fig. 4-22), or sometimes as large pools containing nests of tumor cells. One must be sure that one is looking at tumor hyaluronate and not stroma. Comments made in cases referred to us indicate that some pathologists interpret stromal staining as specific, whereas any tumor stroma will stain to some extent with alcian blue and staining will decrease after treatment with hyaluronidase.

Immunohistochemical Stains

An enormous variety of immunostains have been proposed as adjunctive procedures in the diagnosis of mesothelioma (13,22,88,118,125, 155,183,185). This is an area that is changing rapidly, as new stains and reevaluation of older stains are reported on a monthly basis. Some of the immunohistochemical stains that have withstood the scrutiny of time are listed in Table 4-2. It is important to be aware that different laboratories may obtain dramatically different results with the same antibodies. Thus, proper evaluation of these stains requires the performance of positive and negative controls, and interpretation of the results in the context of all available information.

Broad-spectrum cytokeratin antibodies stain virtually all epithelial mesotheliomas, and do not distinguish mesothelioma from metastatic adenocarcinoma involving the pleura. They do, however, help identify other tumors that can mimic mesothelioma but do not stain for cytokeratin, such as lymphomas, melanomas, and most epithelioid hemangioendotheliomas. Immunostaining for cytokeratin may also aid in the identification of subtle invasion of lung or adipose tissue by mesothelioma not evident on routine H&E-stained sections, permitting its distinction from atypical mesothelial hyperplasia (see chapter 5).

Antibodies to certain subsets of cytokeratins may be useful for the diagnosis of mesothelioma. Most epithelial mesotheliomas react moderately to strongly for antibodies to CK5/6 (see fig. 4-14), whereas most adenocarcinomas do not (37,104,124). However, squamous and large cell lung carcinomas and transitional cell carcinomas of the urinary bladder are frequently positive for CK5/6. Epithelial mesotheliomas typically stain strongly for CK7 but weakly or not at all for CK20 (26). This pattern distinguishes mesothelioma from most gastrointestinal adenocarcinomas, but not from adenocarcinoma of the lung or breast.

Sarcomatous mesotheliomas show a range of broad-spectrum keratin positivity (see fig. 4-34), ranging from virtually 100 percent of cells, to a few scattered positive cells, to, occasionally, no detectable staining. Some areas of a tumor may show intense staining, while adjacent areas may be completely negative. Although the

Table 4-2

IMMUNOHISTOCHEMICAL STAINS USEFUL FOR THE DIAGNOSIS OF MESOTHELIOMA

Stain	Epithelioid Mesothelioma	Expected Result[a] Adenocarcinoma	Sarcomatoid Mesothelioma
Pancytokeratin	Pos	Pos	Pos
Cytokeratin 5/6	Pos	Neg	Neg
Calretinin	Pos[c]	Neg	Neg
Thrombomodulin	Pos[c]	Neg	Neg
WT1[b]	Pos[c]	Neg	Neg
Mesothelin	Pos	Neg	Neg
HBME-1	Pos[d]	Neg	Neg
CEA	Neg	Pos	Neg
BerEP4	Neg	Pos	Neg
CD15 (LeuM1)	Neg	Pos	Neg
B72.3	Neg	Pos	Neg
Bg8	Neg	Pos	Neg
MOC31	Neg	Pos	Neg
TTF1	Neg	Pos[c,e]	Neg

[a]Result observed in most but not necessarily all cases. See text for details.
[b]WT1 = Wilms' tumor antigen 1; CEA = carcinoembryonic antigen; TTF1 = thyroid transcription factor 1.
[c]Nuclear staining.
[d]Surface membrane staining.
[e]Lung and thyroid only.

concept of totally keratin-negative sarcomatoid mesothelioma is somewhat controversial, in the authors' experience a small percentage of otherwise typical cases do not stain with the usual broad-spectrum antikeratin antibodies (7 percent of a series of 280 sarcomatous mesotheliomas [VLR, unpublished data]). The criteria for the diagnosis of keratin-negative sarcomatoid mesothelioma include a typical gross distribution and no evidence of a prior soft tissue sarcoma. Given the plasticity of tumor cells and the vagaries of immunohistochemistry, we believe that such tumors should be considered mesotheliomas.

With these caveats in mind, immunostaining for cytokeratins is useful for distinguishing sarcomatous mesothelioma from most other sarcomas that secondarily involve the pleura. Staining for cytokeratins does not distinguish sarcomatous mesothelioma from pleural synovial sarcomas or sarcomatous lung carcinomas secondarily involving the pleura (19,100), nor does it distinguish between sarcomatous mesothelioma and organizing pleuritis (31; see chapter 5). The identification of keratin-positive spindle cells invading into adipose tissue can aid in the distinction between desmoplastic mesothelioma (see fig. 4-38C) and organizing pleuritis (see chapter 5).

A variety of other markers have been developed that stain most epithelial mesotheliomas but not the majority of carcinomas. Sarcomatous mesotheliomas usually fail to express any of these markers. These include calretinin, Wilms' tumor antigen (WT1), and possibly thrombomodulin, mesothelin, and HBME-1. Calretinin positivity is observed in more than 90 percent of epithelial mesotheliomas (35); both nuclear and cytoplasmic staining may be observed, but nuclear staining is more specific (see fig. 4-14B). Calretinin staining is rare in most adenocarcinomas, but is common in giant cell, small cell, and large cell carcinomas of pulmonary origin and in synovial sarcomas (99,100). Calretinin expression may be observed focally in some sarcomatous mesotheliomas.

Epithelial mesotheliomas stain for WT1 in a nuclear distribution in 70 to 90 percent of cases, whereas such staining is not observed in nonsmall cell carcinomas of lung origin (3,118). WT1 does stain other types of carcinomas, in particular, serous carcinoma of the ovary (118).

Mesothelin has also been suggested as a useful marker for the diagnosis of epithelial mesothelioma. In one series, mesothelin stained 100 percent of epithelial mesotheliomas and nonmucinous ovarian carcinomas that were tested (116). A significant proportion (40 percent) of lung adenocarcinomas, however, were positive and this marker also stained the majority of pancreatic adenocarcinomas, desmoplastic small round cell tumors, and the epithelial component of biphasic synovial sarcoma. Tumors that do not stain for mesothelin include renal cell carcinoma, prostatic adenocarcinoma, hepatocellular carcinoma, thyroid carcinoma, adrenal cortical carcinoma, and carcinoid tumors.

Thrombomodulin staining is controversial; although some authors have reported staining in 80 to 90 percent of epithelial mesotheliomas, typically in a surface membrane distribution (36,122), with only a small percentage of positive lung adenocarcinomas, others have found that both mesotheliomas and carcinomas stain with the same frequency (118). Thrombomodulin stains most squamous cell carcinomas and about 25 percent of small and large cell carcinomas of pulmonary origin (100). Blood vessels are also positive for thrombomodulin, rendering interpretation of tumor cell staining difficult in some cases (36).

HBME-1 is a monoclonal antibody raised against mesothelial cells that stains in a surface membrane distribution. The value of this antibody is also disputed: it has been found to be useful for distinguishing epithelial mesothelioma from adenocarcinoma by some investigators (98,155) but not by others (34,74,114,122).

A number of glycoprotein markers have been developed that stain most adenocarcinomas but fail to stain most epithelial mesotheliomas. These include carcinoembryonic antigen (CEA), BerEP4, CD15 (LeuM1), B72.3, Bg8, and MOC-31 (see figs. 4-14B, 4-56C) (6,22,85,120,127,141, 167,178,185). Adenocarcinomas from different primary sites stain to varying degrees with these markers, as do a small percentage of mesotheliomas. Foci of tumor necrosis and infiltrating inflammatory cells also stain with some of these markers (e.g., CD15 and CEA). The focal nature of the staining may give false negative results in small biopsy specimens. Nonsmall cell lung carcinomas other than adenocarcinoma usually do not react with these antibodies.

Thyroid transcription factor 1 (TTF1) is a useful marker when the differential diagnosis is between epithelial mesothelioma and pulmonary adenocarcinoma (fig. 4-56C). Antibodies for this marker stain more than 70 percent of pulmonary adenocarcinomas in a nuclear pattern (188), but do not stain epithelial mesotheliomas (121). This marker is also positive in thyroid carcinomas, but is negative in most adenocarcinomas from other sites.

Vimentin has been suggested to be a useful marker for epithelial mesothelioma by some authors, but others have found that it also stains a substantial number of pulmonary adenocarcinomas (28,53,105). It is more useful as a marker of adequate fixation, especially in the evaluation of sarcomatous malignancies. Sarcomatous mesotheliomas may stain for S-100 protein, actin, and desmin, in addition to cytokeratins and vimentin. Therefore, these markers cannot be used to distinguish sarcomatous mesothelioma from the spindle cell variant of melanoma (positive for S-100 protein) or leiomyosarcoma (positive for actin and desmin).

A number of additional markers have been reported to be useful in the differential diagnosis of malignant mesothelioma. These include the beta chain of platelet-derived growth factor (136–138), CA19-9 and human placental glycogen (119), CA125 (78), E-cadherin and N-cadherin (131), epidermal growth factor (138), K1 (24), lectins (73), Leu7 and neuron-specific enolase (96), parathyroid hormone–related protein (32), P170 glycoprotein (138), secretory component (43), telomerase reverse transcriptase (TERT) (82), vascular cell adhesion molecule (187), and 44-3A6 (157). None has been shown to be a specific marker for mesothelioma, and more work needs to be done before any of them can be incorporated into routine diagnostic use.

The role of immunohistochemical markers to distinguish malignant from reactive mesothelial lesions is controversial. This topic is discussed in chapter 5.

No single marker is diagnostic of mesothelioma. Consequently, a panel of antibodies is typically used to assist in diagnosis. Each laboratory needs to establish a set of markers that they use. For epithelial mesotheliomas, a good approach is to pick two markers that are typically positive in mesothelioma and two that are typically positive in adenocarcinoma; other markers are then added when needed. With the exception of cytokeratins and perhaps vimentin, immunohistochemical markers are less helpful in the diagnosis of sarcomatous mesothelioma.

The adjunctive nature of immunohistochemical markers must be emphasized. Their use should never be a substitute for careful examination of morphology on H&E-stained sections, correlation with the gross distribution of tumor as determined by the surgeon at the time of thoracoscopy/thoracotomy, and radiographic studies. This comment applies to the other special techniques as well. Attempting to classify a very poorly differentiated pleural tumor purely on the basis of special stains is likely to lead to misdiagnosis.

Ultrastructure

Ultrastructural features may be useful in the diagnosis of tumors of the serosal membranes. The most striking feature of epithelial mesothelioma is the presence of long, often branching, surface microvilli that are devoid of a glycocalyx (fig. 4-51) (18,37,38,128,177). These often have a length to diameter ratio that exceeds 10 to 1 (179, 180). The morphology of the microvilli may be seen particularly well by scanning electron microscopy (67). In contrast, most adenocarcinomas have short, stubby microvilli covered with a glycocalyx and often associated with glycocalyceal bodies (fig. 4-52). Their length to diameter ratio is usually less than 10 to 1. These values, however, are not absolute and there is often overlap in the length to width ratio of mesotheliomas and adenocarcinomas (180). Rootlets at the base of the microvilli favor adenocarcinoma. The presence of microvilli abutting extracellular collagen is a common feature of mesothelioma, but seldom seen in adenocarcinoma.

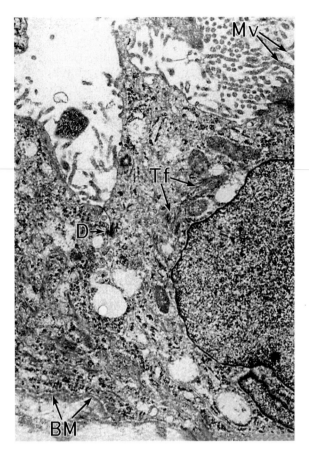

Figure 4-51

ULTRASTRUCTURE OF AN EPITHELIAL MESOTHELIOMA

This electron micrograph shows long, slender microvilli (Mv), tonofibrillar bundles (Tf), desmosomes (D), and basal lamina (BM). (Fig. 15 from Roggli VL, Kolbeck J, Sanfilippo F, Shelburne JD. Pathology of human mesothelioma: etiologic and diagnostic considerations. Pathol Annu 1987;22(Pt 2):108.)

Other ultrastructural features commonly seen in epithelial mesotheliomas include perinuclear tonofibrillar bundles, cytoplasmic glycogen, basal lamina, and long desmosomes (128,143). Mesotheliomas lack lamellar bodies, Clara cell granules, mucous granules, and dense core neurosecretory granules. The findings of pinocytotic vesicles and Weibel-Palade bodies are suggestive of epithelial hemangioendothelioma, while the identification of premelanosomes favors malignant melanoma.

There is no single ultrastructural feature that is the hallmark of mesothelioma, but rather a constellation of findings that may be diagnostically useful. Even definite mesotheliomas tend

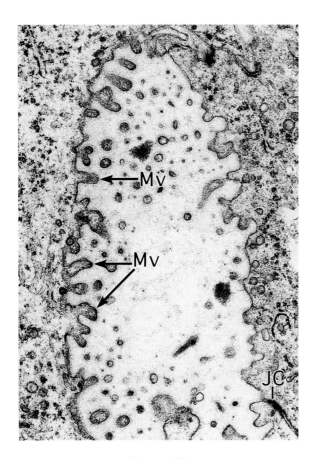

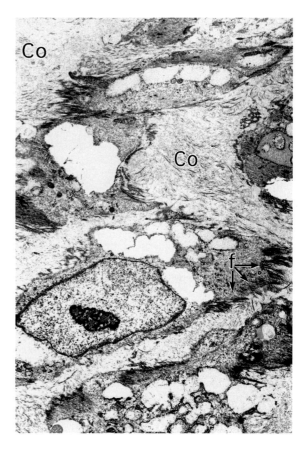

Figure 4-52

**ULTRASTRUCTURE OF A
PULMONARY ADENOCARCINOMA**

Blunt surface microvilli (Mv) and a junctional complex (JC) are observed in this adenocarcinoma metastatic to the pleura. Intermediate filaments and tonofibrillar bundles are not identified. (Courtesy of Dr. S. Hammar, Bremerton, WA.)

Figure 4-53

**ULTRASTRUCTURE OF A
SARCOMATOUS MESOTHELIOMA**

This tumor demonstrates spindle cells with cytoplasmic filaments (f) and abundant extracellular collagen (Co). (Fig. 17 from Roggli VL, Kolbeck J, Sanfilippo F, Shelburne JD. Pathology of human mesothelioma: Etiologic and diagnostic considerations. Pathol Annu 1987;22(Pt 2):110.)

to vary considerably in their ultrastructural features (37). The absence of any or all of these features does not necessarily exclude a diagnosis of mesothelioma, and the interpretation of the ultrastructure should be performed in conjunction with the histologic findings and gross distribution of the tumor. Although the preservation of tissues retrieved from paraffin blocks is often poor, many of the ultrastructural features that are useful for the diagnosis of mesothelioma, such as long surface microvilli and perinuclear tonofilaments, are usually preserved in such specimens (143).

Electron microscopy is less useful for the diagnosis of sarcomatous mesothelioma. These tumors usually resemble fibrosarcomas or myo-

fibroblastic tumors (fig. 4-53), although epithelial differentiation may be seen in some cases. The latter manifests as occasional microvilli, incomplete basal lamina, intercellular junctions, or tonofilaments. Ultrastructural features may be useful for distinguishing sarcomatous mesothelioma from the occasional angiosarcoma, schwannoma, leiomyosarcoma, rhabdomyosarcoma, or spindle cell melanoma that may secondarily involve the pleura (128).

Ultrastructural examination is most useful in epithelial tumors that have fairly clear differentiation by light microscopy. Poorly differentiated tumors tend to have nonspecific ultrastructural features (37,38).

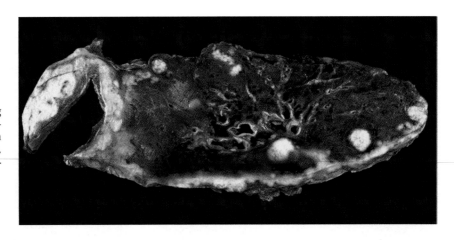

Figure 4-54

CARCINOMA METASTATIC TO THE PLEURA

Primary carcinoma of lung grew around the lung, mimicking a mesothelioma. Most such lesions are adenocarcinomas, but this was a squamous cell carcinoma.

Separation of Mesothelioma from Carcinoma Using Microarrays

Gordon et al. (48a) have recently reported that mesotheliomas and carcinomas can be separated using microarray analysis and the ratio of expression levels of a defined set of genes. Whether this approach will prove to be a useful adjunct to the diagnosis of mesothelioma remains to be established.

DIFFERENTIAL DIAGNOSIS

Tumors Metastatic to the Pleura, Peritoneum, and Pericardium

Virtually any malignant tumor can metastasize to the serous membranes. This section considers only diffuse tumors; localized processes are described in chapters 6 and 7.

In the pleura, data from series of needle biopsies suggest that lung and breast are the most common sites of origin (29). Grossly, pleural metastases can range from multiple small nodules to complete encasement of the lung by tumor. Encasement of the lung by direct spread or metastasis occurs with primary lung adenocarcinomas (40,55,80,112), squamous cell carcinomas (fig. 4-54), small cell carcinomas (45), thymomas (7,65,130), thyroid carcinomas (102), renal cell carcinomas (97), soft tissue sarcomas, melanomas (54), and lymphomas (fig. 4-55), as well as a variety of other sarcomas (see below). Such spread has been labeled *pseudomesotheliomatous* (44,55). Many of the primary lung cancers that grow in this fashion are peripheral adenocarcinomas, and in some instances, radiographic examination reveals a peripheral coin lesion. In others, the exact site of the primary tumor is not obvious (40,55,80,112).

The same processes occur in the peritoneal cavity. Metastatic carcinomas of the ovary, gastrointestinal tract, and less commonly, lung, breast, and uterus, as well as sarcomas, can produce a pattern of multiple small tumor nodules or, occasionally, encase viscera. A similar pattern is seen with primary serous papillary tumors of the peritoneum (see below).

Microscopically, carcinomas and sarcomas metastatic to the serosal membranes usually look like their primaries, but, particularly in the pleura, a desmoplastic reaction may occur which mimics the response seen with mesotheliomas (fig. 4-56A,B). The morphologic features of carcinoma and mesothelioma have been compared above, and in most instances the distinction is obvious. Immunohistochemical staining, as described in the section Special Techniques, separates most carcinomas from mesotheliomas (fig 4-56C), although some very high-grade lesions cannot be easily classified. Immunohistochemical stains also are useful for separating metastatic sarcomas from mesotheliomas, since the latter are usually broad-spectrum keratin positive and sometimes positive for calretinin and CK5/6 as well.

Primary Serous Papillary Carcinoma/ Serous Papillary Borderline Tumor of the Peritoneum and Metastatic Serous Carcinoma/ Serous Borderline Tumor of the Ovary

Approximately 350 examples of primary serous papillary carcinoma (PSPC) have been reported (152), but these are actually fairly

Figure 4-55

LYMPHOMA METASTATIC TO THE PLEURA

This lymphoma mimics a mesothelioma.

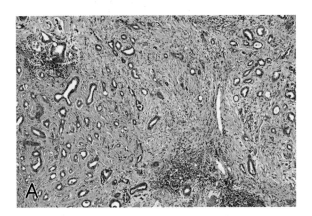
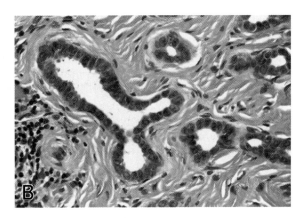

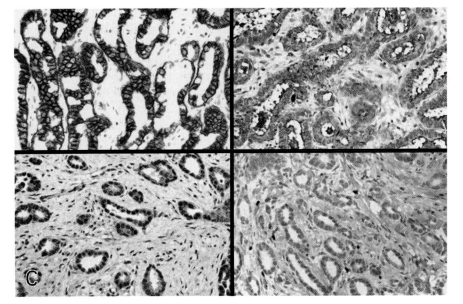

Figure 4-56

ADENOCARCINOMA OF LUNG METASTATIC TO THE PLEURA

A: The pattern of tumor growth, with outward branches and a desmoplastic response, mimics a mesothelioma.

B: Higher-power view. The columnar shape of the cells indicates a carcinoma and not a mesothelioma.

C: The tumor cells are positive for BerEP4 (upper left), CEA (upper right), and TTF1 (lower left), and negative for calretinin (lower right), a typical carcinoma staining pattern.

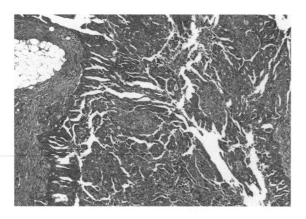 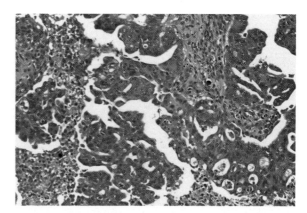

Figure 4-57

PRIMARY SEROUS PAPILLARY CARCINOMA OF THE PERITONEUM

Left: The complex papillations are not a typical feature of mesothelioma. (Courtesy of Dr. P. Clement, Vancouver, BC, Canada.)

Right: Higher-power view of the tumor shows a complex pattern of gland-within-gland growth, typical of carcinomas and not mesotheliomas.

common tumors. PSPCs usually form multiple small nodules over the peritoneal surfaces (172) and thus cannot be distinguished grossly from early peritoneal mesothelioma. A papillary architecture may be visible in larger deposits.

Microscopically, PSPCs and serous borderline tumors typically have elongated, moderately anaplastic cells with cellular stratification (fig. 4-57). This is different from the usual appearance of mesothelioma, which has much more monotonous cells and usually only a single cell layer covering the papillary structures. Psammoma bodies, which occur but are uncommon in mesotheliomas (see fig. 4-20), are frequent in PSPC. In some tumors, psammoma bodies are so numerous that the term *psammocarcinoma* has been suggested. While serous tumors can produce papillae with fibrovascular cores and sometimes only a single layer of covering cells, structures that mimic the appearance of mesotheliomas, they do not form the outwardly branching glands that are common in mesotheliomas. Necrosis may be found in PSPC (172), but this feature is unusual in untreated mesotheliomas, especially epithelial forms. The appearance of metastatic PSPC and borderline tumor is identical. A small number of papillary serous carcinomas of the paratesticular region have also been reported (69) and must be considered in the differential diagnosis of mesothelioma of the tunica vaginalis.

In most instances, the diagnosis of PSPC is readily apparent with routine stains. Histochemical and immunohistochemical separation of serous tumors from mesotheliomas is a problem that has not been entirely satisfactorily resolved. A proportion of serous tumors, but not all, produce dPAS-positive mucin droplets (152), and there is considerable controversy concerning the proportion of these tumors that stain for typical carcinoma markers such as CEA, LeuM1, B72.3, or BerEP4 (9,120,172). Truong et al. (172) found that 85 percent of their cases were positive for B72.3. Ordonez (120) concluded that calretinin, thrombomodulin, and CK5/6 were good mesothelial markers in this setting; MOC-31, B72.3, BerEP4, CA19-9 and LeuM1 stained PSPCs; and WT1 stained both (118). A recent study reported that nuclear staining for calretinin was only seen in the cells of epithelial peritoneal mesotheliomas and not in PSPC, whereas BerREP4 staining was seen in most serous papillary tumors and very few mesotheliomas (9). Staining for thrombomodulin, CK5/6, and CD44H (hyaluronate receptor) showed considerable overlap, as did that for CA125. Other authors have reported that BerEP4 staining is common in mesotheliomas (117), although in our experience such staining is uncommon. At this point, calretinin, B72.3, and BerEP4 appear to be the most useful stains for separating mesotheliomas from PSPC and borderline tumors.

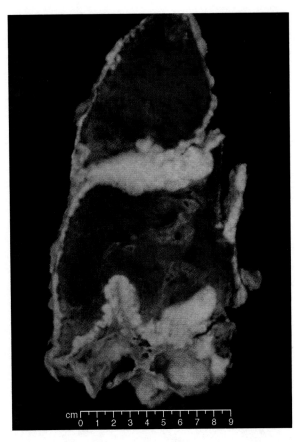

Figure 4-58

PRIMARY HEMANGIOENDOTHELIOMA OF THE PLEURA
The gross appearance mimics that of a mesothelioma.

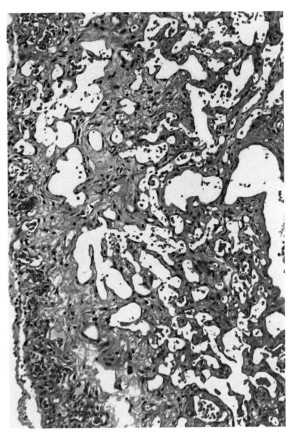

Figure 4-59

PRIMARY ANGIOSARCOMA OF THE PLEURA
This tumor is relatively well differentiated.

Primary Squamous Carcinomas of the Pleura

A small number of tumors claimed to be primary squamous carcinomas of the pleura have been reported in patients with chronic pleural inflammation, almost all after tuberculous empyema or plombage therapy (108,147). It is unclear whether these are true squamous carcinomas or mesotheliomas with squamous differentiation, an uncommon but well-recognized phenomenon.

Primary Sarcomas of the Pleura

Vascular Sarcomas. Rare examples of primary angiosarcomas, epithelioid hemangioendotheliomas, and lymphangiosarcomas of the pleura, pericardium, and peritoneum have been reported (8,44,89,90,160,191). Grossly, they spread diffusely over the serosal surfaces and can form a rind that mimics a malignant mesothe-

lioma (fig. 4-58). Microscopically, these tumors resemble their soft tissue counterparts. In some instances, a fairly well-differentiated angiosarcomatous pattern is observed (fig. 4-59). Primary hemangioendotheliomas appear to be more common than angiosarcomas. They form a variety of abortive capillaries and larger vascular spaces, occasionally with papillary formations (figs. 4-60, 4-61). The cells covering the papillae, however, are endothelial and not mesothelial. Some tumors have a spindle cell component.

The plump cells of epithelioid hemangioendothelioma can sometimes be confused with mesothelial cells. Indeed, one of the original terms used for mesothelioma was endothelioma. Immunohistochemical staining is extremely helpful in arriving at the correct diagnosis. Vascular sarcomas are sometimes weakly positive for broad-spectrum keratin (88), but they always stain for one or more of CD31 (figs. 4-60, right; 4-61,

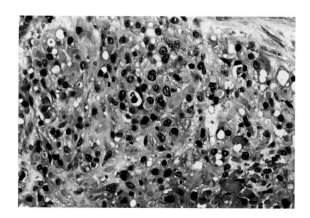
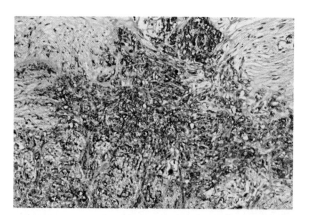

Figure 4-60

PRIMARY HEMANGIOENDOTHELIOMA OF THE PLEURA

Left: This field resembles an epithelial mesothelioma, although the tumor is cytologically high grade.
Right: Diffuse CD31 staining.

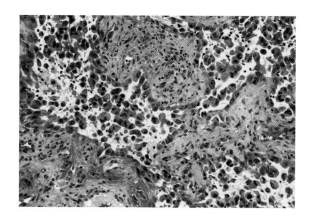
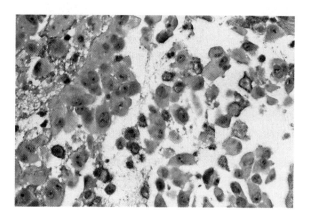

Figure 4-61

PRIMARY HEMANGIOENDOTHELIOMA OF THE PLEURA

Left: The polygonal cells resemble mesothelial cells.
Right: The tumor stains for CD31.

right), CD34, factor VIII, and *Ulex europaeus* agglutinin I, whereas malignant mesotheliomas do not (8,88). All the reported tumors have been highly aggressive lesions leading to the rapid death of the patient.

Synovial Sarcoma. A small number of primary biphasic and monophasic synovial sarcomas have been described in the pleural cavity (10,49). The patients are on the whole much younger than the typical mesothelioma patient, with almost all cases occurring before the age of 50. The sex ratio is approximately 1 to 1. Most patients have a pleural effusion and all have a mass lesion visible radiographically. Many synovial sarcomas are localized lesions and some are pedunculated, but others show diffuse pleural spread.

Microscopically, the spindle cell component of pleural synovial sarcoma is composed of cells that are generally much smaller and that possess much less cytoplasm than the typical cells of sarcomatous mesothelioma (fig. 4-62). Some tumors have a hemangiopericytoma-like pattern of growth. In biphasic tumors, the epithelial component usually forms cleft-like spaces, a phenomenon that is unusual in the epithelial component of malignant mesothelioma.

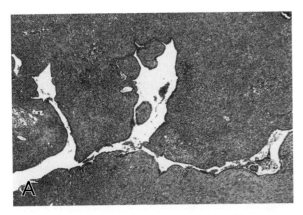

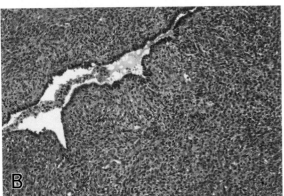

Figure 4-62

PRIMARY SYNOVIAL SARCOMA OF PLEURA

A: The combination of small spindled cells and cleft-like spaces with cuboidal to slightly columnar epithelial lining is typical of biphasic synovial sarcoma.

B: Higher-power view shows very small spindled cells, a feature common in synovial sarcoma but not in mesothelioma.

C: Digested periodic acid–Schiff (dPAS) stain demonstrates secretion of PAS-positive material.

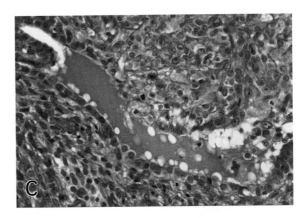

The epithelial component of the biphasic tumor usually secretes material into the epithelial lumens that stains with dPAS and mucicarmine (fig. 4-62C). The epithelial component is keratin positive, and even the monophasic forms usually show focal keratin positivity. Calretinin positivity is common in synovial sarcomas (100) and thus is not helpful for differentiation from mesothelioma. Many synovial sarcomas are BerEP4 positive (49,100), a feature that has been proposed as a useful discriminator from malignant mesothelioma, although, as noted above, there is considerable controversy in the literature about the frequency with which mesotheliomas stain for BerEP4 (117). Synovial sarcomas also frequently stain for bcl-2 (62,166). Reverse transcriptase polymerase chain reaction (RT-PCR) can be used to show the presence of a *SYT-SSX1* or *SYT-SSX2* fusion transcript, a reasonably specific diagnostic finding in synovial sarcomas (10).

The prognosis of patients with pleural synovial sarcomas, at least the localized ones, appears to be considerably better than that for patients with malignant mesotheliomas. About half of the 14 patients described by Aubry et al.

(10) were alive without evidence of disease, one for 8 years. The diffuse forms, however, are always highly aggressive tumors.

Other Primary Sarcomas. Primary liposarcomas occasionally occur in the pleura, but the reported cases all appear to be localized rather than diffuse lesions (115,186). A malignant peripheral nerve sheath tumor simulating a mesothelioma has been described (126); we have also seen an example that occurred in a patient with neurofibromatosis. Tumors believed to be primary osteosarcomas and chondrosarcomas have been reported, but, given the propensity of mesotheliomas to show chondrosarcomatous and osteosarcomatous differentiation, it is not clear whether a diagnosis of primary osteosarcoma or chondrosarcoma is possible. Although some authors have claimed that primary malignant fibrous histiocytomas occur in the pleura (108), current practice classifies diffuse tumors with the appearance of fibrosarcoma or malignant fibrous histiocytoma in the pleura or peritoneum as sarcomatous mesothelioma. A positive stain for broad-spectrum keratin in any of these tumors supports the diagnosis of malignant mesothelioma, but, since some sarcomatous mesotheliomas are keratin negative, a negative result does not rule out a mesothelioma.

Miscellaneous Tumors. Thymomas may reach the pleura as metastatic lesions, but primary pleural thymomas also occur (fig. 4-63). Epithelial thymoma can spread along the pleural surface and mimic mesothelioma (7,65,130). Attention to the histologic pattern seen on routinely stained specimens is crucial to making the diagnosis, since these lesions are typically keratin, calretinin, and CK5/6 positive. However, they may show CD20 positivity (7). The presence of a lobular architecture separated by broad fibrous bands and perivascular spaces is useful diagnostically for thymoma.

CHOICES AND LIMITATIONS OF DIAGNOSTIC SPECIMENS

Tissue Specimens

As a rule, thoracoscopic/laparoscopic or open surgical biopsies are needed for the diagnosis of malignant mesothelioma. Needle biopsies occasionally allow a definitive diagnosis, but are generally not as useful as larger specimens. Boutin and Rey (16) reported a specific diagnosis rate of 98 percent for thoracoscopy specimens, 26 percent for cytology specimens, and only 21 percent for needle biopsy specimens in a series of 188 patients. Malignant mesotheliomas of epithelial type may be very bland cytologically, whereas reactive mesothelial cells may mimic malignancy because of cytologic atypia, presence of mitoses, and architectural features (see chapter 5). Proliferating fibroblasts may mimic malignant mesotheliomas of sarcomatous type. In these settings, a sufficiently large tissue sample is necessary to confirm tissue invasion, an important criterion of malignancy with mesothelial proliferations (see chapter 5). Desmoplastic mesotheliomas are particularly difficult to diagnose with small samples that do not permit evaluation of stromal invasion or show sarcomatous areas of the tumor. Sometimes desmoplastic mesotheliomas can only be diagnosed on large, pleural stripping specimens.

The problems of a cytology-based diagnosis of epithelial mesothelioma have been discussed in chapter 3. Cytology specimens from serosal effusions caused by sarcomatous mesothelioma

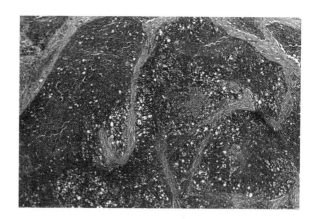

Figure 4-63

PRIMARY PLEURAL THYMOMA
(Courtesy of Dr. T. Colby, Scottsdale, AZ.)

or desmoplastic mesothelioma are often negative for malignant cells because these types of mesothelioma do not shed cells into the serosal effusion. The clinical impression should be considered in rendering a diagnosis, particularly when based on a small biopsy. If the process is clinically malignant and the biopsy unequivocally benign, the biopsy may have missed a diagnostic lesion.

STAGING, THERAPY, AND PROGNOSIS

There is no universally agreed upon method of staging malignant mesotheliomas, in part because the prognosis has been so uniformly dire that staging seems to serve little purpose. The most detailed staging system is that proposed by the International Mesothelioma Interest Group (Table 4-3) (149), but other, simpler, systems are also currently employed (Table 4-4) (164). It has been claimed that the International Mesothelioma Interest Group staging does not predict survival (192). Some of the staging schemes do show a correlation with survival: the staging system proposed by Sugarbaker et al. (164) and shown in Table 4-4, shows median survival periods of 25, 20, and 16 months for patients with stage 1, 2, or 3 tumors, respectively, who were treated with extrapleural pneumonectomy and adjuvant chemotherapy. These survival figures are, unfortunately, somewhat misleading, since most patients present at high stage of disease and are not candidates for extrapleural pneumonectomy. In

Table 4-3

STAGING SYSTEM PROPOSED BY THE INTERNATIONAL MESOTHELIOMA INTEREST GROUP[a]

Tumor (T) Staging

T1a	Tumor limited to the ipsilateral parietal pleura, including the mediastinal and diaphragmatic pleura, without involvement of the visceral pleura
T1b	Tla + Scattered foci of tumor involving the visceral pleura
T2	Tumor involving each of the ipsilateral pleural surfaces (parietal, mediastinal, diaphragmatic, and visceral pleura) with at least one of the following features:
	Involvement of diaphragmatic muscle
	Confluent visceral pleural tumor (including the fissures) or extension of tumor from the visceral pleura into the underlying pulmonary parenchyma
T3	Locally advanced but potentially resectable tumor. The tumor involves all of the ipsilateral pleural surfaces with at least one of the following features:
	Involvement of the endothoracic fascia
	Extension into the mediastinal fat
	A solitary, completely resectable focus of tumor extending into the soft tissues of the chest wall
	Nontransmural involvement of the pericardium
T4	Locally advanced, technically unresectable tumor. The tumor involves all of the ipsilateral pleural surfaces with at least one of the following features:
	Diffuse extension or metastatic spread to the chest wall with or without rib destruction
	Direct trans-diaphragmatic extension to the peritoneum
	Direct extension to the contralateral pleura
	Direct extension to any mediastinal organ
	Direct extension to the spine

Lymph Node (N) Staging

NX	Regional lymph nodes (LNs) cannot be assessed
N0	No regional LN metastases
N1	Involvement of ipsilateral bronchopulmonary or hilar LNs
N2	Involvement of subcarinal or ipsilateral mediastinal LNs (including the internal mammary LNs)
N3	Involvement of the contralateral mediastinal or internal mammary LNs or any supraclavicular LNs

Metastases (M) Staging

MX	Presence of distant metastases cannot be assessed
M0	No distant metastases
M1	Distant metastases present

Overall Staging

Stage I		
	Ia	T1a N0 M0
	Ib	T1b N0 M0
Stage II		T2 N0 M0
Stage III		Any T3 M0
		Any N1 M0
		Any N2 M0
Stage IV		Any T4
		Any N3
		Any M1

[a]Data from reference 149.

seven published series constituting approximately 3,400 patients diagnosed over 30 years, the median survival period ranged from about 6 to 12 months (2,59,86,140,149,159,188).

Surgery, radiotherapy, or chemotherapy alone appears to have no effect on survival, although local radiotherapy may prevent chest wall and needle tract seeding (192). More recently, trimodality therapy, consisting of extrapleural pneumonectomy with en-bloc resection of the parietal pleura, lung, pericardium, and diaphragm, followed by radiation and chemotherapy, has been employed in a selected set of patients with low-stage tumors. For 176 patients who received this treatment and survived the immediate postsurgical period, the 2- and 5-year survival rates were 38 and 15 percent (164). For those patients with epithelial tumors and

Table 4-4

STAGING SYSTEM PROPOSED BY SUGARBAKER ET AL.[a]

Stage	Definition
1	Disease completely resected within the capsule of the parietal pleura without adenopathy; ipsilateral pleura, lung, pericardium, diaphragm, or chest wall disease limited to previous biopsy sites
2	All of stage 1 with positive resection margins and/or intrapleural adenopathy
3	Local extension into the chest wall or mediastinum, into the heart or through the diaphragm or peritoneum; or with extrapleural lymph node involvement
4	Distant metastatic disease

[a]Data from reference 164.

Sugarbaker postoperative stage 1 disease (negative margins and negative lymph nodes), the 2- and 5-year survival rates were 68 and 46 percent (164).

A number of factors have been shown to influence prognosis: worse prognosis is associated with an older age at diagnosis and the presence of chest pain, dyspnea, and weight loss (5,59, 139,162). The histologic pattern is also important: it has been consistently found that patients whose tumors show a purely epithelial histology fare best; those with a purely sarcomatous histology, worst; and those with mixed epithelial and sarcomatous patterns are intermediate between the two (2,5,20,42,59,139,148,188), although the differences in median survival periods in large series are only a matter of a few months. In the series of patients treated with trimodality therapy described above, no patient with a sarcomatous tumor survived for 5 years (192). Gordon et al. (51) recently reported that relative gene expression levels, as determined by microarray analysis, might be useful in predicting tumor behavior. This approach needs considerable further verification and in particular needs to be applied to purely epithelial tumors.

As opposed to pleural tumors, there is increasing evidence that some peritoneal tumors, particularly in women, are curable. Kerrigan et al. (76) reported that, in a series of 25 female patients, 40 percent survived more than 4 years and none of these patients died of disease; the longest survival period at last follow-up was 15 years. Histologic appearance did not predict survival. Similarly, Sugarbaker et al. (164a) reported that, in a series of 68 patients treated with tumor debulking and intraperitoneal chemotherapy, the median survival period was 67 months.

Because of the poor prognosis and relatively high stage of tumor seen in most patients with mesothelioma, a variety of new and experimental therapies have been proposed, including use of angiogenesis inhibitors, epidermal growth factor–receptor inhibitors, platelet-derived growth factor inhibitors, suicide gene therapy with a herpes simplex tyrosine kinase gene and subsequent treatment with the antiviral agent gancyclovir, and immunomodulatory gene therapy by transfection of a variety of cytokines such as IL-2, IL-6, and IL-1β (113). At this point it is unclear if any of these approaches will be successful.

REFERENCES

1. Aberle DR. High-resolution computed tomography of asbestos-related disease. Semin Roentgenol 1991;26:118–31.

2. Alberts AS, Falkson G, Goedhals L, Vorobiof DA, Van der Merwe CA. Malignant pleural mesothelioma: a disease unaffected by current therapeutic maneuvers. J Clin Oncol 1988;6:527–35.

3. Amin KM, Litzky LA, Smythe WR, et al. Wilms' tumor 1 susceptibility (WT1) gene products are selectively expressed in malignant mesothelioma. Am J Pathol 1995;146:344–56.

4. Amin MB, Tamboli P, Merchant SH, et al. Micropapillary component in lung adenocarcinoma: a distinctive histological feature with possible prognostic significance. Amer J Surg Path 2002;26:358–64.

5. Antman K, Shemin R, Ryan L, et al. Malignant mesothelioma: prognostic variables in a registry of 180 patients, the Dana-Farber Cancer Institute and Brigham and Woman's Hospital experience over two decades, 1965-1985. J Clin Oncol 1988;6:147–53.

6. Arber DA, Weiss LM. CD15: a review. Appl Immunohistochem 1993;1:17–30.

7. Attanoos RL, Galateau-Salle F, Gibbs AR, Muller S, Ghandour F, Dojcinov SD. Primary thymic epithelial tumours of the pleura mimicking malignant mesothelioma. Histopathology 2002;41:42–9.

8. Attanoos RL, Suvarna SK, Rhead E, et al. Malignant vascular tumours of the pleura in "asbestos" workers and endothelial differentiation in malignant mesothelioma. Thorax 2000;55:860–3.

9. Attanoos RL, Webb R, Dojcinov SD, Gibbs AR. Value of mesothelial and epithelial antibodies in distinguishing diffuse peritoneal mesothelioma in females from serous papillary carcinoma of the ovary and peritoneum. Histopathology 2002;40:237–44.

10. Aubry MC, Bridge JA, Wickert R, Tazelaar HD. Primary monophasic synovial sarcoma of the pleura: five cases confirmed by the presence of SYT-SSX fusion transcript. Amer J Surg Path 2001;25:776–81.

11. Baris YI, Artvinli M, Sahin AA. Environmental mesothelioma in Turkey. Ann N Y Acad Sci 1979;330:423–32.

12. Battifora H, McCaughey WT. Tumors of the serosal membranes. Atlas of Tumor Pathology, 3rd Series, Fascicle 15. Washington, DC: Armed Forces Institute of Pathology; 1995.

13. Bedrossian CW, Bonsib S, Moran C. Differential diagnosis between mesothelioma and adenocarcinoma: a multimodal approach based on ultrastructure and immunocytochemistry. Semin Diagn Pathol 1992;9:124–40.

14. Bolen JW, Hammar SP, McNutt MA. Reactive and neoplastic serosal tissue. A light-microscopic, ultrastructural, and immunocytochemical study. Am J Surg Pathol 1986;10:34–47.

15. Bolen JW, Hammar SP, McNutt MA. Serosal tissue: reactive tissue as a model for understanding mesotheliomas. Ultrastruct Pathol 1987;11:251–62.

16. Boutin C, Rey F. Thoracoscopy in pleural malignant mesothelioma: a prospective study of 188 consecutive patients. Part 1: Diagnosis. Cancer 1993;72:389–93.

17. Browne K. Asbestos-related disorders. In: Parkes WR, ed. Occupational lung disorders, 3rd ed. 1994:411–504.

18. Burns TR, Greenberg SD, Mace ML, Johnson EH. Ultrastructural diagnosis of epithelial malignant mesothelioma. Cancer 1985;56:2036–40.

19. Cagle PT, Truong LD, Roggli VL, Greenberg SD. Immunohistochemical differentiation of sarcomatoid mesotheliomas from other spindle cell neoplasms. Am J Clin Pathol 1989;92:566–71.

20. Cantin R, Al-Jabi M, McCaughey WT. Desmoplastic diffuse mesothelioma. Am J Surg Pathol 1982;6:215–22.

21. Carbone M, Kratzke RA, Testa JR. The pathogenesis of mesothelioma. Semin Oncol 2002;29:2–17.

22. Carella R, Deleopnardi G, D'Errico A, et al. Immunohistochemical panels for differentiating epithelial malignant mesothelioma from lung adenocarcinoma: a study with logistic regression analysis. Amer J Surg Pathol 2001;25:43–50.

23. Cavazza A, Travis LB, Travis WD, et al. Post-irradiation malignant mesothelioma. Cancer 1996;77:1379–85.

24. Chang K, Pai LH, Pass H, et al. Monoclonal antibody K1 reacts with epithelial mesothelioma but not with lung adenocarcinoma. Am J Surg Pathol 1992;16:259–68.

25. Cheng JQ, Lee WC, Klein MA, Cheng GZ, Jhanwar SC, Testa JR. Frequent mutations of NF2 and allelic loss from chromosome band 22q12 in malignant mesothelioma: evidence for a two-hit mechanism of NF2 inactivation. Genes Chromosomes Cancer 1999;24:238–42.

26. Chu P, Wu E, Weiss LM. Cytokeratin 7 and cytokeratin 20 expression in epithelial neoplasms: a survey of 435 cases. Mod Pathol 2000; 13:962–72.

27. Churg A. Deposition and clearance of chrysotile asbestos. Ann Occup Hyg 1994;38:625–34.

28. Churg A. Immunohistochemical staining for vimentin and keratin in malignant mesothelioma. Am J Surg Pathol 1985;9:360–5.

29. Churg A. Lung biopsy, lung resection, and autopsy lung specimens: handling and diagnostic limitations. In: Churg A, Myers J, Tazelaar H, Wright JL, eds. Thurlbeck's pathology of the lung, 3rd ed. New York: Thieme; 2005:95–108.

30. Churg A. Neoplastic asbestos-induced disease. In: Churg A, Green FH, eds. Pathology of occupational lung disease, 2nd ed. Baltimore: Williams & Wilkins; 1998:339–92.

31. Churg A, Colby TV, Cagle P, et al. The separation of benign and malignant mesothelial proliferations. Amer J Surg Pathol 2000;24:1183–200.

32. Clark SP, Chou ST, Martin TJ, Danks JA. Parathyroid hormone-related protein antigen localization distinguishes between mesothelioma and adenocarcinoma of lung. J Pathol 1995;176:161–5.

33. Clement PB, Young RH, Scully RE. Malignant mesotheliomas presenting as ovarian masses. A report of nine cases, including two primary ovarian mesotheliomas. Am J Surg Pathol 1996;20:1067–80.

34. Comin CE, Novelli L, Boddi V, Paglierani M, Dini S. Calretinin, thrombomodulin, CEA, and CD15: a useful combination of immunohistochemical markers for differentiating pleural epithelial mesothelioma from peripheral pulmonary adenocarcinoma. Hum Pathol 2001;32:529–36.

35. Cury PM, Butcher DN, Corrin B, Nicholson AG. The use of histological and immunohistochemical markers to distinguish pleural malignant mesothelioma and in situ meosthelioma from reactive mesothelial hyperplasia and reactive pleural fibrosis. J Pathol 1999;189:251–7.

36. Cury PM, Butcher DN, Fisher C, Corrin B, Nicholoson AG. Value of the mesothelium-associated antibodies thrombomodulin, cytokeratin 5/6, calretinin, and CD44H in distinguishing epithelial pleural mesothelioma from adenocarcinoma metastatic to the pleura. Mod Pathol 2000;13:107–12.

37. Dardick I, Al-Jabi M, McCaughey WT, Deodhare S, van Nostrand AW, Srigley JR. Diffuse epithelial mesothelioma: a review of the ultrastructural spectrum. Ultrastruct Pathol 1987;11:503–33.

38. Dardick I, Al-Jabi M, McCaughey WT, Srigley JR, van Nostrand AW, Ritchie AC. Ultrastructure of poorly differentiated diffuse epithelial mesotheliomas. Ultrastruct Pathol 1984;7:151–60.

39. De Rienzo A, Testa JR. Recent advances in the molecular analysis of human malignant mesothelioma. Clin Ter 2000;151:433–8.

40. Dessy E, Pietra GG. Pseudomesotheliomatous carcinoma of the lung. An immunohistochemical and ultrastructural study of three cases. Cancer 1991;68:1747–53.

41. Ellis K, Wolff M. Mesotheliomas and secondary tumors of the pleura. Semin Roentgenol 1977;l2:303–12.

42. Elmes PC, Simpson JC. The clinical aspects of mesothelioma. Quar J Med 1976;45:427–48.

43. Ernst CS, Brooks JJ. Immunoperoxidase localization of secretory component in reactive mesothelium and mesotheliomas. J Histochem Cytochem 1981;29:1102–4.

44. Falconieri G, Bussani R, Mirra M, Zanella M. Pseudomesotheliomatous angiosarcoma: a pleuropulmonary lesion simulating malignant pleural mesothelioma. Histopathology 1997;30:419–24.

45. Falconieri G, Zanconati F, Bussani R, Di Bonito L. Small cell carcinoma of lung simulating pleural mesothelioma. Report of 4 cases with autopsy confirmation. Path Res Pract 1995;191:1147–52.

46. Foddis R, De Rienzo A, Broccoli D, et al. SV40 infection induces telomerase activity in human mesothelial cells. Oncogene 2002;221:1434–42.

47. Fraire AE, Greenberg SD, Spjut HJ, et al. Effect of fibrous glass on rat pleural mesothelium. Histopathologic observations. Am J Respir Crit Care Med 1994;150:521–7.

48. Fraire AE, Greenberg SD, Spjut HJ, et al. Effect of erionite on the pleural mesothelium of the Fischer 344 rat. Chest 1997;111:1375–80.

49. Gaertner E, Zeren EH, Fleming MV, Colby TV, Travis WD. Biphasic synovial sarcoma arising in the pleural cavity. A clinicopathologic study of five cases. Amer J Surg Pathol 1996;20:36–45.

50. Gordon GJ, Jensen RV, Hsiao LL, et al. Translation of microarray data into clinically relevant cancer diagnostic tests using gene expression ratios in lung cancer and mesothelioma. Can Res 2002;62:4963–7.

51. Gordon GJ, Jensen RV, Hsiao LL, et al. Using gene expression ratios to predict outcome among patients with mesothelioma. J Natl Cancer Inst 2003;95:598–605.

52. Grundy GW, Miller RW. Malignant mesothelioma in childhood. Report of 13 cases. Cancer 1972;30:1216–8.

53. Hammar SP. Pleural diseases. In: Dail DH, Hammar SP, eds. Pulmonary pathology, 2nd ed. New York: Springer-Verlag; 1994:1463–579.

54. Hammar SP, Bockus DE, Remington FL, Rohrbach KA. Mucin-positive epithelial mesotheliomas: a histochemical, immunohistochemical, and ultrastructural comparison with mucin-producing pulmonary adenocarcinomas. Ultrastruct Pathol 1996;20:293–325.

55. Harwood TR, Gracey DR, Yokoo H. Pseudomesotheliomatous carcinoma of the lung. A variant of peripheral lung cancer. Am J Clin Pathol 1976;65:159–67.

56. Henderson DW, Attwood HD, Constance TJ, Shilkin KB, Steele RH. Lymphohistiocytoid mesothelioma: a rare lymphomatoid variant of predominantly sarcomatoid mesothelioma. Ultrastruct Pathol 1988;12:367–84.

57. Henderson DW, Shilkin KB, Langlois SL, Whitaker D, eds. Malignant mesothelioma. New York: Hemisphere Pub Corp; 1992.

58. Henderson DW, Shilkin KB, Whitaker D. Reactive mesothelial hyperplasia vs mesothelioma, including mesothelioma in situ: a brief review. Am J Clin Pathol 1998;110:397–404.

59. Herndon JE, Green MR, Chahinian AP, Corson JM, Suzuki Y, Vogelzang NJ. Factors predictive of survival among 337 patients with mesothelioma treated between 1984 and 1994 by the Cancer and Leukemia Group B. Chest 1998;113: 723–31.

60. Higashihara M, Sunaga S, Tange T, Oohashi H, Kurokawa K. Increased secretion of interleukin-6 in malignant mesothelioma cells from a patient with marked thrombocytosis. Cancer 1992;70:2105–8.

61. Hillerdal G, Berg J. Malignant mesothelioma secondary to chronic inflammation and old scars. Two new cases and review of the literature. Cancer 1985;55:1968–72.

62. Hirakawa N, Naka T, Yamamoto I, Fukuda T, Tsuneyoshi M. Overexpression of bcl-2 protein in synovial sarcoma: a comparative study of other soft tissue spindle cell sarcomas and additional analysis by fluorescence in situ hybridization. Hum Pathol 1996;27:1060–5.

63. Hirao T, Bueno R, Chen CJ, Gordon GJ, Heilig E, Kelsey KT. Alterations of the p16(INK4) locus in human malignant mesothelial tumors. Carcinogenesis 2002;23:1127–30.

64. Hodgson JT, Darnton A. The quantitative risks of mesothelioma and lung cancer in relation to asbestos exposure. Ann Occup Hyg 2000;44: 565–601.

65. Hofmann W, Moller P, Manke HG, Otto HF. Thymoma. A clinicopathologic study of 98 cases with special reference to three unusual cases. Pathol Res Pract 1985;179:337–53.

66. Jagirdar J, Frydman C, Sakurai H, Dumitrescu O. Mesothelial papillary proliferation of the pleura associated with radiation therapy: does it have a role in the pathogenesis of mesothelioma? Mt. Sinai J Med 1989;56:147–9.

67. Jandik WR, Landas SK, Bray CK, Lager DJ. Scanning electron microscopic distinction of pleural mesotheliomas from adenocarcinomas. Mod Pathol 1993;6:761–4.

68. Jones MA, Young RH, Scully RE. Malignant mesothelioma of the tunica vaginalis. A clinicopathologic analysis of 11 cases with review of the literature. Am J Surg Pathol 1995;19:815–25.

69. Jones MA, Young RH, Srigley JR, Scully RE. Paratesticular serous papillary carcinoma. A report of six cases. Am J Surg Pathol 1995;19: 1359–65.

70. Kannerstein M, Churg J. Peritoneal mesothelioma. Hum Pathol 1977;8:83–93.

71. Kannerstein M, Churg J. Desmoplastic diffuse malignant mesothelioma. Prog Surg Pathol 1980;1:19–27.

72. Kaplan MA, Tazelaar HD, Hayashi T, Schroer KR, Travis WD. Adenomatoid tumors of the pleura. Am J Surg Pathol 1996;20:1219–23.

73. Kawai T, Greenberg SD, Truong LD, Mattioli CA, Klima M, Titus ML. Differences in lectin-binding of malignant pleural mesothelioma and adenocarcinoma of the lung. Am J Pathol 1988;130: 401–10.

74. Kennedy AD, King G, Kerr KM. HBME-1 and antithrombomodulin in the differential diagnosis of malignant mesothelioma of pleura. J Clin Pathol 1997;50:859–62.

75. Kerrigan SA, Cagle P, Churg A. Malignant mesothelioma of the peritoneum presenting as an inflammatory lesion: a report of four cases. Am J Surg Pathol 2003;27:248–53.

76. Kerrigan SA, Turnnir RT, Clement PE, Young RH, Churg A. Diffuse malignant mesotheliomas of the peritoneum in women: a clinicopathologic study of 25 patients. Cancer 2002;94:378–85.

77. Khalidi HS, Medeiros LJ, Battifora H. Lymphohistiocytoid mesothelioma. An often misdiagnosed variant of sarcomatoid malignant mesothelioma. Am J Clin Pathol 2000;113:649–54.

78. Khoury N, Raju U, Crissman JD, Zarbo RJ, Greenwald KA. A comparative immunohistochemical study of peritoneal and ovarian serous tumors, and mesothelioma. Hum Pathol 1990;21:811–9.

79. Kitamura F, Araki S, Suzuki Y, Yokoymana K, Tanigawa T, Iwasaki R. Assessment of the mutations of p53 suppressor gene and Ha- and Ki-ras oncogenes in malignant mesothelioma in relation to asbestos exposure: a study of 12 American patients. Ind Health 2002;40:175–81.

80. Koss M, Travis W, Moran C, Hochholzer L. Pseudomesotheliomatous adenocarcinoma: a reappraisal. Sem Diag Pathol 1992;9:117–23.

81. Krismann M, Muller KM, Jaworska M, Johnen G. Molecular cytogenetic differences between histological subtypes of malignant mesotheliomas: DNA cytometry and comparative genomic hybridization of 90 cases. J Pathol 2002;197: 363–71.

82. Kumaki F, Kawai T, Churg A, et al. Expression of telomerase reverse transcriptase (TERT) in malignant mesotheliomas. Am J Surg Pathol 2002;26:365–70.

83. La Vecchia C, Decarlia A, Peto J, Levi F, Tomei F, Negri E. An age, period and cohort analysis of pleural cancer mortality in Europe. Eur J Cancer Prev 2000;9:179–84.

84. Lanphear BP, Buncher CR. Latent period for malignant mesothelioma of occupational origin. J Occup Med 1992;34:718–21.

85. Latza U, Niedobitek G, Schwarting R, Nekarda H Stein H. Ber-Ep4: new monoclonal antibody which distinguishes epithelia from mesothelia. J Clin Pathol 1990;43:213–9.

86. Legha SS, Muggia FM. Pleural mesothelioma: clinical features and therapeutic implications. Ann Intern Med 1977;87:613–20.

87. Leigh J, Davison P, Hendrie L, Berry D. Malignant mesothelioma in Australia, 1945-2000. Am J Ind Med 2002;41:188–201.

88. Leong AS, Vernon-Roberts E. The immunohistochemistry of malignant mesothelioma. Pathol Annu 1994;29(Pt 2):157–79.

89. Lin BT, Colby T, Gown AM, et al. Malignant vascular tumors of the serous membranes mimicking mesothelioma. A report of 14 cases. Am J Surg Path 1996;20:1431–9.

90. McCaughey WT, Dardick I, Barr JR. Angiosarcoma of serous membranes. Arch Path Lab Med 1983;107:304–7.

91. McCaughey WT, Kannerstein M, Churg J. Tumors and pseudotumors of the serous membranes. Atlas of Tumor Pathology, 2nd Series, Fascicle 20. Washington, DC: Armed Forces Institute of Pathology; 1985.

92. McDonald JC, McDonald AD. Epidemiology of asbestos-related lung cancer. In: Antman K, Aisner J, eds. Asbestos-related malignancy. Orlando: Grune & Stratton; 1987:57–79.

93. Mangano WE, Cagle PT, Churg A, Vollmer RT, Roggli VL. The diagnosis of desmoplastic malignant mesothelioma and its distinction from fibrous pleurisy: a histologic and immunohistochemical analysis of 31 cases including p53 immunostaining. Am J Clin Pathol 1998;110: 191–9.

94. Marom EM, Erasmus JJ, Pass HI, Patz EF Jr. The role of imaging in malignant pleural mesothelioma. Semin Oncol 2002;29:26–35.

95. Mayall FG, Gibbs AR. The histology and immunochemistry of small cell mesothelioma. Histopathology 1992;20:47–51.

96. Mayall FG, Jasani B, Gibbs AR. Immunohistochemical positivity for neuron-specific enolase and Leu-7 in malignant mesotheliomas. J Pathol 1991;165:325–8.

97. McCaughey WT, Colby TV, Battifora H, et al. Diagnosis of diffuse malignant mesothelioma: experience of a US/Canadian Mesothelioma Panel. Mod Pathol 1991;4:342–53.

98. Miettinen M, Kovatich AJ. HBME-1: a monoclonal antibody useful in the differential diagnosis of mesothelioma, adenocarcinoma, and soft-tissue and bone tumors. Appl Immunohistochem 1995;3:115–22.

99. Miettinen M, Limon J, Niezabitowski A, Lasota J. Calretinin and other mesothelioma markers in synovial sarcoma: analysis of antigenic similarities and differences with malignant mesothelioma. Am J Surg Pathol 2001;25:610–7.

100. Miettinen M, Sarlomo-Rikala M. Expression of calretinin, thrombomodulin, keratin 5, and mesothelin in lung carcinomas of different types: an immunohistochemical analysis of 596 tumors in comparison with epithelioid mesotheliomas of the pleura. Am J Surg Pathol 2003;27:150–8.

101. Miyamoto Y, Nakano S, Shimazaki Y, Matsuda H, Fukuda H. Pericardial mesothelioma presenting as left atrial thrombus in a patient with mitral stenosis. Cardiovasc Surg 1996;4:51–2.

102. Mizukami Y, Michigishi T, Nonomura A, et al. Distant metastases in differentiated thyroid carcinomas: a clinical and pathologic study. Hum Pathol 1990;21:283–90.

103. Moertel CG. Peritoneal mesothelioma. Gastroenterology 1972;63:346–50.

104. Moll R, Dhouailly D, Sun TT. Expression of keratin 5 as a distinctive feature of epithelial and biphasic mesotheliomas: an immunohistochemical study using monoclonal antibody AE14. Virchows Arch B Cell Pathol Incl Mol Pathol 1989;58:129–45.

105. Mullink H, Henzen-Longmans SC, Alons-van Kordelaar JJ, Tadema TM, Meijer CJ. Simultaneous immunoenzyme staining of vimentin and cytokeratins with monoclonal antibodies as an aid in differential diagnosis of malignant mesothelioma from pulmonary adenocarcinoma. Virchows Arch B Cell Pathol Incl Mol Pathol 1986;52:55–65.

106. Murthy SS, Testa JR. Asbestos, chromosomal deletions, and tumor suppressor gene alterations in human malignant mesothelioma. J Cell Physiol 1999;180:150–7.

107. Musk AW, Dewar J, Shilkin KB, Whitaker D. Miliary spread of malignant pleural mesothelioma without a clinically identifiable pleural tumour. Aust N Z J Med 1991;21:460–2.

108. Myoui A, Aozasa K, Iuchi K, et al. Soft tissue sarcoma of the pleural cavity. Cancer 1991;68:1550–4.

109. Neugat AI, Ahsan H, Antman KH. Incidence of pleural mesothelioma after thoracic radiotherapy. Cancer 1997;80:948–50.

110. Ni Z, Liu Y, Keshava N, Zhou G, Whong W, Ong T. Analysis of K-ras and p53 mutations in mesotheliomas from humans and rats exposed to asbestos. Mutat Res 2000;468:87–92.

111. Nicholson CP, Donohue JH, Thompson GB, Lewis JE. A study of metastatic cancer found during inguinal hernia repair. Cancer 1992;69: 3008–11.

112. Nishimoto Y, Ohno T, Saito K. Pseudomesotheliomatous carcinoma of the lung with histochemical and immunohistochemical study. Acta Pathol Jap 1983;33:415–23.

113. Nowak AK, Lake RA, Kindler HL, Robinson BW. New approaches for mesothelioma: biologics, vaccines, gene therapy, and other novel agents. Semin Oncol 2002;29:82–96.

114. Oates J, Edwards C. HBME-1, MOC-31, WT1 and calretinin: an assessment of recently described markers for mesothelioma and adenocarcinoma. Histopathology 2000;36:341–7.

115. Okby NT, Travis WD. Liposarcoma of the pleural cavity: clinical and pathologic features of 4 cases with a review of the literature. Arch Pathol Lab Med 2000;124:699–703.

116. Ordonez NG. Application of mesothelin immunostaining in tumor diagnosis. Am J Surg Pathol 2003;27:1418–28.

117. Ordonez NG. The immunohistochemical diagnosis of epithelial mesothelioma. Hum Pathol 1999;30:313–23.

118. Ordonez NG. Immunohistochemical diagnosis of epithelioid mesotheliomas: a critical review of old markers, new markers. Hum Pathol 2002;33:953–67.

119. Ordonez NG. The immunohistochemical diagnosis of mesothelioma. Differentiation of mesothelioma and lung adenocarcinoma. Am J Surg Pathol 1989;13:276–91.

120. Ordonez NG. Role of immunochemistry in distinguishing epithelial peritoneal mesotheliomas from peritoneal and ovarian serous carcinomas. Am J Surg Pathol 1998;22:1203–14.

121. Ordonez NG. Thyroid transcription factor-1 is a marker of lung and thyroid carcinomas. Adv Anat Pathol 2000;7:123–7.

122. Ordonez NG. The value of antibodies 44-3A6, SM3, HBME-1, and thrombomodulin in differentiating epithelial pleural mesothelioma from lung adenocarcinoma: a comparative study with other commonly used antibodies. Am J Surg Pathol 1997;21:1399–408.

123. Ordonez NG. Value of the Ber-EP4 antibody in differentiating epithelial pleural mesothelioma from adenocarcinoma. The M.D. Anderson experience and a critical review of the literature. Am J Clin Pathol 1998;109:85–9.

124. Ordonez NG. Value of cytokeratin 5/6 immunostaining in distinguishing epithelial mesothelioma of the pleura from lung adenocarcinoma. Am J Surg Pathol 1998;22:1215–21.

125. Ordonez NG, Mackay B. The roles of immunohistochemistry and electron microscopy in distinguishing epithelial mesothelioma of the pleura from adenocarcinoma. Adv Anat Pathol 1996;3:273–93.

126. Ordonez NG, Tornos C. Malignant peripheral nerve sheath tumor of the pleura with epithelial and rhabdomyoblastic differentiation: report of a case clinically simulating mesothelioma. Am J Surg Pathol 1997;21:1515–21.

127. Otis CN, Carter D, Cole S, Battifora H. Immunohistochemical evaluation of pleural mesothelioma and pulmonary adenocarcinoma. A bi-institutional study of 47 cases. Am J Surg Pathol 1987;11:445–56.

128. Oury TD, Hammar SP, Roggli VL. Ultrastructural features of diffuse malignant mesotheliomas. Hum Pathol 1998;29:1382–92.

129. Papp T, Schipper H, Pemsel H, et al. Mutational analysis of N-ras, p53, p16INK4a, p14ARF and CDK4 genes in primary human malignant mesothelioma. Int J Oncol 2001;18:425–33.

130. Payne CB Jr, Morningstar WA, Chester EH. Thymoma of the pleura masquerading as diffuse mesothelioma. Am Rev Respir Dis 1966;94:441–46.

131. Peralta Soler A, Knudsen KA, Jaurand MC, et al. The differential expression of N-cadherin and E-cadherin distinguishes pleural mesotheliomas from lung adenocarcinomas. Hum Pathol 1995;26:1363–9.

132. Perez-Ordonez B, Srigley JR. Mesothelial lesions of the paratesticular region. Semin Diagn Pathol 2000;17:294–306.

133. Perks WH, Crow JC, Green M. Mesothelioma associated with the syndrome of inappropriate secretion of antidiuretic hormone. Am Rev Respir Dis 1978;117:789–94.

134. Plas E, Riedl CR, Pfulger H. Malignant mesothelioma of the tunica vaginalis testis: review of the literature and assessment of prognostic parameters. Cancer 1998;83:2437–46.

135. Price B. Analysis of current trends in United States mesothelioma incidence. Am J Epidemiol 1997;145:211–8.

136. Ramael M, Buysse C, van den Bossche J, Segers K, van Marck E. Immunoreactivity for the beta chain of the platelet-derived growth factor receptor in malignant mesothelioma and non-neoplastic mesothelium. J Pathol 1992;167:1–4.

137. Ramael M, Segers K, Buysse C, Van den Bossche J, Van Mark E. Immunohistochemical distribution patterns of epidermal growth factor receptor in malignant mesothelioma and non-neoplastic mesothelium. Virchows Arch A Pathol Anat Histopathol 1991;419:171–5.

138. Ramael M, van den Bossche J, Buysse C, et al. Immunoreactivity for p-170 glycoprotein in malignant mesothelioma and in non-neoplastic mesothelium of the pleura using the murine monoclonal antibody JSB-1. J Pathol 1992;167:5–8.

139. Ribak J, Selikoff IJ. Survival of asbestos insulation workers with mesothelioma. Br J Ind Med 1992;49:732–5.

140. Rich S, Presant CA, Meyer J, Stevens SC, Carr D. Human chorionic gonadotropin and malignant mesothelioma. Cancer 1979;43:1457–62.

141. Riera JR, Astengo-Osuna C, Longmate JA, Battifora H. The immunohistochemical diagnostic panel for epithelial mesothelioma: a re-evaluation after heat-induced epitope retrieval. Am J Surg Pathol 1997;21:1409–19.

142. Roberts GH. Distant visceral metastases in pleural mesothelioma. Br J Dis Chest1976;70:246–50.

143. Roggli VL, Kolbeck J, Sanfilippo F, Shelburne JD. Pathology of human mesothelioma. Etiologic and diagnostic considerations. Pathol Annu 1987;22(Pt 2):91–131.

144. Roggli VR, Sanfillipo F, Shelburne JD. Mesothelioma. In: Roggli VL, Greenberg SD, Pratt PC. Pathology of asbestos-associated diseases. Boston: Little, Brown; 1992:109–264.

145. Roggli VL, Sharma A, Butnor KJ, Sporn T, Vollmer RT. Malignant mesothelioma and occupational exposure to asbestos: a clinicopathological correlation of 1445 cases. Ultrastruct Pathol 2002;26:55–65.

146. Ros PR, Yuschok TJ, Buck JL, Shekitka KM, Kaude JV. Peritoneal mesothelioma. Radiologic appearances correlated with histology. Acta Radiol. 1991;32:355–8.

147. Roviaro GC, Sartori F, Calabro F, Varoli F. The association of pleural mesothelioma and tuberculosis. Am Rev Respir Dis 1982;126:569–71.

148. Ruffie P, Feld R, Minkin S, et al. Diffuse malignant mesothelioma of the pleura in Ontario and Quebec: a retrospective study of 332 patients. J Clin Oncol 1989;7:1157–68.

149. Rusch VW. A proposed new international TMN staging system for malignant pleural mesothelioma. From the International Mesothelioma Interest Group. Chest 1995;108:1122–8.

150. Sandhu H, Dehnen W, Roller M, Abel J, Unfried K. mRNA expression in patterns in different stages of asbestos-induced carcinogenesis in rats. Carcinogenesis 2000;21:1023–9.

151. Scharnhorst V, van der Eb AJ, Jochemsen AG. WT1 proteins: functions in growth and differentiation. Gene 2001;273:141–61.

152. Scully RE, Young RH, Clement PB. Tumors of the ovary, maldeveloped gonads, fallopian tube, and broad ligament. Atlas of Tumor Pathology, 3rd Series, Fascicle 23. Washington, DC: Armed Forces Institute of Pathology; 1998.

153. Selikoff IJ, Hammond EC, Seidman H. Latency of asbestos disease among insulation workers in the United States and Canada. Cancer 1980; 46:2736–40.

154. Serio G, Scattone A, Pennella A, et al. Malignant deciduoid mesothelioma of the pleura. Histopathology 2002;40:348–52.

155. Sheibani K, Esteban JM, Bailey A, Battifora H, Weiss LM. Immunopathologic and molecular studies as an aid to the diagnosis of malignant mesothelioma. Hum Pathol 1992;23:107–16.

156. Situnayake RD, Middleton WG. Recurrent pneumothorax and malignant pleural mesothelioma. Respir Med 1991;85:255–6.

157. Spagnolo DV, Whitaker D, Carrello S, Radosevich JA, Rosen ST, Gould VE. The use of monoclonal antibody 44–3A6 in cell blocks in the diagnosis of lung carcinoma, carcinomas metastatic to lung and pleura, and pleural malignant mesothelioma. Am J Clin Pathol 1991;95:322–9.

158. Spirtas R, Beebe GW, Connelly RR, et al. Recent trends in mesothelioma incidence in the United States. Am J Indust Med 1986;9:397–407.

159. Spirtas R, Heineman EF, Bernstein L, et al. Malignant mesothelioma: attributable risk of asbestos exposure. Occup Environ Med 1994;51:804–11.

160. Sporn TA, Butnor KJ, Roggli VL. Epithelioid hemangioendothelioma of the pleura: an aggressive vascular malignancy and clinical mimic of malignant mesothelioma. Histolopathology 2002;41(Suppl 2):173–7.

161. Stedman's Medical Dictionary, 22nd ed. Baltimore: Williams & Wilkins; 1972:425, 1119.

162. Steele JP. Prognostic factors in mesothelioma. Sem Oncol 2002;29:36–40.

162a. Stout AP. Tumors of the pleura. Harlem Hosp Bull 1971;5:54–7.

163. Strickler HD, Goedert JJ, Devesa SS, Lahey J, Fraumeni JF Jr, Rosenberg PF. Trends in US pleural mesothelioma incidence rates following simian virus 40 contamination of early poliovirus vaccines. J Natl Cancer Inst 2003;95:38–45.

164. Sugarbaker DJ, Flores R, Jacklitsch MT, et al. Resection margins, extrapleural nodal status, and cell type determine postoperative long-term survival in trimodality therapy of malignant pleural mesothelioma: results in 183 patients. J Thorac Cardiovasc Surg 1999;117:54–65.

164a. Sugarbaker PH, Welch LS, Mohamed F, Glehen O. A review of peritoneal mesothelioma at the Washington Cancer Institute. Surg Oncol Clin N Am 2003;12:605–21.

165. Sussman J, Rosai J. Lymph node metastasis as the initial manifestation of malignant mesothelioma. Report of six cases. Am J Surg Pathol 1990;14:819–28.

166. Suster S, Fisher C, Moran CA. Expression of bcl-2 oncoprotein in benign and malignant spindle cell tumors of soft tissue, skin, serosal surfaces, and gastrointestinal tract. Am J Surg Pathol 1998;22:863–72.

167. Szpak CA, Johnston WW, Roggli V, et al. The diagnostic distinction between malignant mesothelioma of the pleura and adenocarcinoma of the lung as defined by a monoclonal antibody (B72.3). Am J Pathol 1986;122:252–60.

168. Talerman A, Montero JR, Chilcote RR, Okagaki T. Diffuse malignant peritoneal mesothelioma in a 13-year-old girl. Report of a case and review of the literature. Am J Surg Pathol 1985;9:73–80.

169. Testa JR, Giordano A. SV40 and cell cycle perturbations in malignant mesothelioma. Semin Cancer Biol 2001;11:31–8.

170. Thomason R, Schlegel W, Lucca M, Cummings S, Lee S. Primary malignant mesothelioma of the pericardium. Case report and literature review. Tex Heart Inst J 1994;21:170–4.

171. Travis WD, Colby TV, Corrin B, et al. Histological typing of lung and pleural tumours, 3rd ed. Berlin: Springer; 1999:23.

172. Truong LD, Maccato ML, Awalt H, Cagle PT, Schwartz MR, Kaplan AL. Serous surface carcinoma of the peritoneum: a clinicopathologic study of 22 cases. Hum Pathol 1990;21:99–110.

173. Turk J, Kenda M, Kranjec I. Primary malignant pericardial mesothelioma. Klin Wochenschr 1991;69:674–8.

174. Vasilieva LA, Pylev LN, Rovensky YA. Pathogenesis of experimentally induced asbestos mesothelioma in rats. Cancer Lett 1998;134:209–16.

175. Wagner JC, Sleggs C, Marchand P. Diffuse pleural mesothelioma and asbestos exposure in the North Western Cape Province. Br J Ind Med 1960;17:260–71.

176. Wagner JC, Newhouse ML, Corrin B, Rossiter CE, Griffiths DM. Correlation between fibre content of the lung and disease in east London asbestos factory workers. Br J Ind Med 1988;45:305–8.

177. Wang NS. Electron microscopy in the diagnosis of pleural mesotheliomas. Cancer 1973;31:1046–54.

178. Wang NS, Huang SN, Gold P. Absence of carcinoembryonic antigen-like material in mesothelioma: an immunohistochemical differentiation from other lung cancers. Cancer 1979;44:937–43.

179. Warhol MJ, Corson JM. An ultrastructural comparison of mesotheliomas with adenocarcinomas of the lung and breast. Hum Pathol 1985;16:50–5.

180. Warhol MJ, Hickey WF, Corson JM. Malignant mesothelioma: ultrastructural distinction from adenocarcinoma. Am J Surg Pathol 1982;6:307–14.

181. Watanabe A, Sakata J, Kawamura H, Yamada O, Matsuyama T. Primary pericardial mesothelioma presenting as constrictive pericarditis: a case report. Japan Circ J 2000;64:385–8.

182. Whitaker D, Henderson DW, Shilkin KB. The concept of mesothelioma in situ: implications for diagnosis and histogenesis. Semin Diagn Pathol 1992;9:151–61.

183. Wick MR, Loy T, Mills SE, Leigier JF, Manivel JC. Malignant epithelioid pleural mesothelioma versus peripheral pulmonary adenocarcinoma: a histochemical, ultrastructural, and immunohistologic study of 103 cases. Hum Pathol 1990;21:759–66.

184. Wilson GE, Hasleton PS, Chatterjee AK. Desmoplastic malignant mesothelioma: a review of 17 cases. J Clin Pathol. 1992;45:295–8.

185. Wirth PR, Legier J, Wright GL Jr. Immunohistochemical evaluation of seven monoclonal antibodies for differentiation of pleural mesothelioma from lung adenocarcinoma. Cancer 1991;67:655–62.

186. Wong WW, Pluth JR, Grado GL, Schild SE, Sanderson DR. Liposarcoma of the pleura. Mayo Clin Proc 1994;69:882–5.
187. Yamada T, Jiping J, Endo R, Gotoh M, Shimosato Y, Hirohashi S. Molecular cloning of a cell-surface glycoprotein that can potentially discriminate mesothelium from epithelium: its identification as vascular cell adhesion molecule 1. Br J Cancer 1995;71:562–75.
188. Yatabe Y, Mitsudomi T, Takahashi T. TTF-1 expression in pulmonary adenocarcinomas. Am J Surg Pathol 2002;26:767–73.
189. Yates DH, Corrin B, Stidolph PN, Browne K. Malignant mesothelioma in south east England: clinicopathologic experience of 272 cases. Thorax 1997;52:507–12.
190. Yousem SA, Hochholzer L. Malignant mesotheliomas with osseous and cartilaginous differentiation. Arch Pathol Lab Med 1987;111:62–6.
191. Yousem SA, Hochholzer L. Unusual thoracic manifestations of epithelioid hemangioendothelioma. Arch Pathol Lab Med 1987;111:459–63.
192. Zellos LS, Sugarbaker DJ. Multimodality treatment of diffuse malignant pleural mesothelioma. Sem Oncol 2002;29:41–50.

5 SEPARATION OF BENIGN AND MALIGNANT MESOTHELIAL PROLIFERATIONS

INTRODUCTION

When malignant mesothelioma became a topic of interest to pathologists in the 1960s, the major issues concerned recognizing mesotheliomas as such, and separating them from tumors metastatic to the pleura and peritoneum. Publications from the 1960s and 1970s were devoted to detailed morphologic descriptions of mesotheliomas. This emphasis has continued, quite appropriately: it can be seen in the Second (13) and the Third Series Fascicles (1), published in 1985 and 1995, and it is evident in this present volume as well. With time, pathologists have learned to recognize the typical patterns of mesothelioma, and the use of immunohistochemistry has helped in separating mesotheliomas from other malignancies, so that identification of a particular tumor as a mesothelioma has become considerably easier.

What has emerged as an increasing problem in the pathology of the serosal membranes is the separation of benign from malignant mesothelial processes. This has proven to be a particularly difficult issue; for example, of 217 cases circulated to the entire United States-Canadian Mesothelioma Reference Panel between 1994 and 1999, a disagreement of some degree (i.e., one or more Panelists disagreeing with the majority) on the question of benign versus malignant was recorded in 22 percent of cases (4). This is a very biased figure because only the difficult cases are circulated to the whole Panel, but it gives some idea of the magnitude of the problem. In most respects, the distinction between benign and malignant is much more crucial to the patient than the question of whether an obviously malignant tumor is a carcinoma or mesothelioma, since, with uncommon exceptions, neither of the latter forms are currently treatable.

This chapter presents our approach to making the distinction between benign and malignant mesothelial proliferations (Table 5-1). We advise the reader to go through chapter 4 first, because a working familiarity with the appearances of clearly defined mesotheliomas is crucial to the issue of what is malignant and what is not. It is also important to have clinical information when making a diagnosis in these cases. Reactive mesothelial processes are commonly associated with effusions and thickening of the affected serosal membranes, as are mesotheliomas, but most mesotheliomas look like malignancies on thoracoscopic or thoracotomy (laparoscopic/ laparotomy) examination and appear as multiple tumor nodules or sheets ("cakes") of tumor on a serosal surface. If the clinician says the process appears benign after surgical or endoscopic examination, or the lesion is described as an inflammatory mass (particularly in the female pelvis), the pathologist should think twice before calling it malignant. Overall, we advise a cautious approach to diagnosis. When there is doubt about whether a mesothelial proliferation is really malignant, we believe it is far preferable for patient management to apply the term "atypical mesothelial proliferation" or "atypical mesothelial hyperplasia." More tissue can usually be obtained if the process is thought to be malignant, and most mesotheliomas make themselves evident in short order.

It is also important to be sure that one is dealing with a mesothelial proliferation; occasionally, carcinomas metastatic to the serosal membranes grow along the serosal surfaces and mimic mesothelial cells. As described in chapter 4, carcinoma cells and mesothelial cells usually look quite different, but if there is doubt, then application of a fairly simple set of immunochemical stains (chapter 4) will usually solve the problem. We are assuming in the following discussion that the cells in question are definitely of mesothelial origin.

As was noted in chapter 4, the presence or absence of a history of asbestos exposure should not be considered in arriving at a diagnosis. Asbestos exposure may cause both mesotheliomas

Table 5-1

SEPARATION OF REACTIVE FROM MALIGNANT MESOTHELIAL PROLIFERATIONS

Benign Mesothelial Reactions	Malignant Mesothelial Neoplasms
No true invasion of stroma (but superficial entrapment may be present in areas of organization)	Invasion of stroma (the deeper, the more definitive)
May be densely cellular in the pleural space, but not in the stroma	Dense cellularity (noninflammatory) in stroma favors malignancy
Process becomes more fibrotic toward chest wall ("zonation")	No zonation to process; often more cellular away from effusion
Cytologic atypia confined to area of organizing effusion	Cytologic atypia present in any area, but many mesotheliomas are deceptively bland and relatively monotonous
Necrosis rare	Necrosis usually a sign of malignancy
Mitoses may be plentiful	Many mesotheliomas show very few mitoses (but atypical mitoses favor malignancy)
No nodular expansion of stroma	Nodular expansions of stroma sometimes present
Storiform pattern absent or minimal	Extensive storiform pattern favors malignancy (desmoplastic mesothelioma) but is not sufficient by itself (see Table 5-2)
Benign reactions may be keratin, p53, and EMA[a] positive	Mesotheliomas are almost always keratin and often EMA and p53 positive

[a]EMA = epithelial membrane antigen.

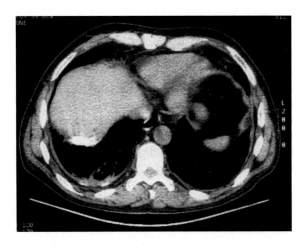

Figure 5-1

BENIGN ASBESTOS-INDUCED PLEURAL FIBROSIS
Computerized tomography (CT) scan.

and benign mesothelial reactions (fig. 5-1), and both conditions may occur in the absence of asbestos exposure.

EPITHELIAL-TYPE PROLIFERATIONS

It is convenient to divide mesothelial proliferations into those in which the cell of interest is of epithelial type and those in which the cell of interest is a spindle cell. Both benign and malignant proliferations often contain a mixture of epithelial and spindle cells, however.

Proliferations Confined to a Serosal Surface

In many instances a biopsy specimen shows proliferating mesothelial cells only on the pleural/peritoneal surface, without any mesothelial cells visible in the underlying tissue. Such surface cells may form a single layer or appear as heaped up aggregates.

Simple Hyperplasia. The normal mesothelium consists of flat, very inconspicuous cells covering the serosal surfaces (fig. 5-2). In simple hyperplasia, the mesothelial cells instead appear as a layer of distinctly prominent, flattened to cuboidal cells arrayed along a serosal membrane. By definition, in simple hyperplasia the cells are relatively bland and monotonous, but sometimes distinct nucleoli may be present (fig. 5-2). The hyperplastic cells are usually regularly spaced along the membrane. Simple hyperplasia may be seen in any process that leads to irritation of a serosal surface; it is very common in (benign) ascites (fig. 5-3); in patients with endometriosis, pelvic inflammatory disease, or ovarian tumors; in hernia sacs (fig. 5-2) and hydroceles; and as part of the pleural reaction to pneumothorax.

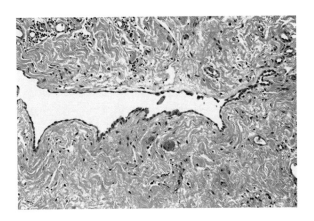
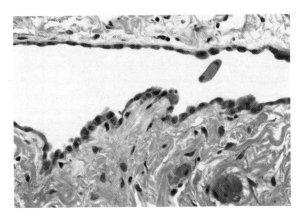

Figure 5-2

SIMPLE MESOTHELIAL HYPERPLASIA IN A HERNIA SAC

Low-power (left) and high-power (right) microscopic views. Normal mesothelial cells are almost flat. Varying degrees of increased cell size and the development of a more cuboidal appearance are typical of simple mesothelial hyperplasia. Nucleoli can be discerned in some of the larger cells in the high-power view.

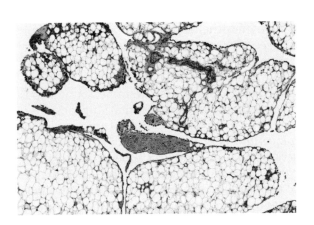
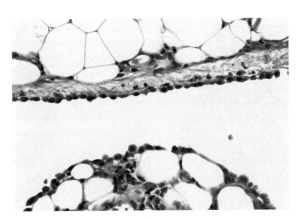

Figure 5-3

SIMPLE MESOTHELIAL HYPERPLASIA IN A PATIENT WITH ASCITES SECONDARY TO CIRRHOSIS

Low-power (left) and high-power (right) microscopic views. The enlarged mesothelial cells are similar to each other, the usual finding in simple hyperplasia.

Atypical Mesothelial Hyperplasia and the Concept of Mesothelioma in Situ. More florid mesothelial proliferations on a serosal surface can take a variety of forms, including single cells with various degrees of cytologic atypia (fig. 5-4), or heaped up aggregates of cells on the surface, usually without papillary cores or any other clear morphologic pattern. The single cells are most often cuboidal, but in atypical reactions more elongated forms with nuclei located more toward the serosal cavity are seen. The cells of atypical mesothelial hyperplasia generally demonstrate prominent nucleoli and sometimes distinctly

enlarged nuclei. There can be considerable variation from cell to cell (fig. 5-5).

There is, unfortunately, no sharp morphologic cutoff between simple hyperplasia and atypical hyperplasia, and the same processes that lead to one can lead to the other. The pathologist needs to make a judgement about whether a given lesion is most likely reactive, most likely malignant (see below), or is indeterminate for malignancy. When the changes in the mesothelial cells are minimal, this determination is fairly easy, but it is important to bear in mind the lesson from exfoliative cytology of effusions:

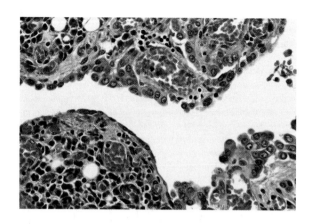
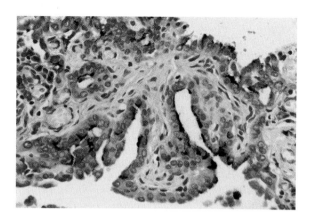

Figure 5-4

ATYPICAL MESOTHELIAL HYPERPLASIA IN THE PERITONEAL CAVITY

The patient had metastatic serous carcinoma of the ovary.

Left: The process, although cytologically atypical, is benign.

Right: The mesothelial cells are epithelial membrane antigen (EMA) positive, illustrating the fact that EMA positivity is not a sign of malignancy.

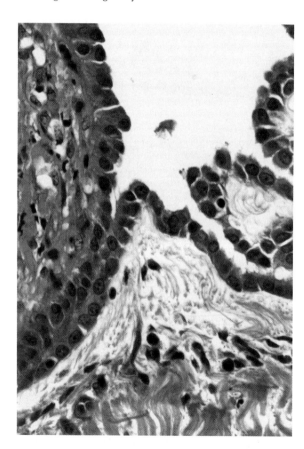

Figure 5-5

**ATYPICAL HYPERPLASIA OF
THE PLEURAL MESOTHELIUM**

Specimen from a young adult with benign spontaneous pneumothorax.

benign mesothelial reactions can be very atypical (see chapter 3). The presence of other inflammatory changes in the biopsy favors a reactive lesion, as does a clinical history that suggests benign disease (for example, spontaneous pneumothorax in a young adult) (fig. 5-5). Hernia sacs, which may be sites of inflammation, frequently show some degree of mesothelial reaction and sometimes this appears cytologically worrisome (16). Conversely, the combination of atypical surface cells and underlying tissue showing no reaction at all, or densely fibrotic underlying tissue, at least raises the possibility of malignancy. If the process does not appear to be inflammatory/reactive and there are no good (benign) reasons to expect a mesothelial reaction, the lesion should be diagnosed as atypical mesothelial hyperplasia or atypical mesothelial proliferation, accompanied by a comment requesting additional tissue if the process is clinically suspicious.

Given the general principles of the morphologic stages leading to malignancy in other organs, one would assume that there exists an entity of "mesothelioma in situ," meaning histologically malignant mesothelial cells arrayed along a serosal surface, but without evidence of invasive tumor. Indeed, Whitaker et al. (19) and Henderson et al. (8) proposed criteria for a morphologic diagnosis of mesothelioma in situ: a single row of highly atypical mesothelial cells that are quite pleomorphic, sometimes produc-

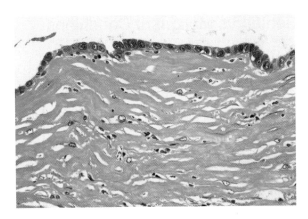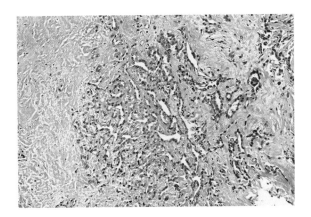

Figure 5-6

ATYPICAL MESOTHELIAL HYPERPLASIA AND INVASIVE MESOTHELIOMA

Left: While the surface proliferation is probably malignant, it should still, by itself, be diagnosed as atypical mesothelial hyperplasia.

Right: Invasive tumor in the same case.

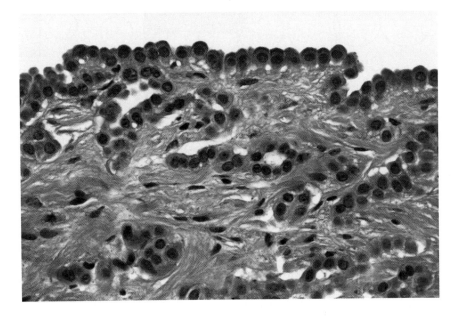

Figure 5-7

ATYPICAL SURFACE HYPERPLASIA OVERLYING INVASIVE MESOTHELIOMA

The surface proliferation is undoubtedly part of the malignant mesothelioma, but by itself cannot be diagnosed as malignant.

ing a picket fence appearance, and that individually appear cytologically malignant. Whitaker only made a diagnosis of mesothelioma in situ when actual nodules of invasive tumor were also present.

There is no doubt that purely surface proliferations of cytologically disturbing cells that presumably represent mesothelioma in situ can sometimes be found away from foci of unequivocal invasive mesothelioma (fig. 5-6). Surface involvement, however, may also take the form of remarkably bland-appearing cells that

are nonetheless clearly part of an underlying mesothelioma (fig. 5-7). Unfortunately, cytologically atypical mesothelial surface reactions may be seen in a variety of clearly benign processes (fig. 5-5) (2) and in processes of doubtful malignancy. We believe that the separation of benign from malignant in this setting using histologic criteria is not feasible. Indeed, the same group that described mesothelioma in situ in the presence of invasive tumor later noted that in the one instance in which they made a diagnosis of mesothelioma in situ without the

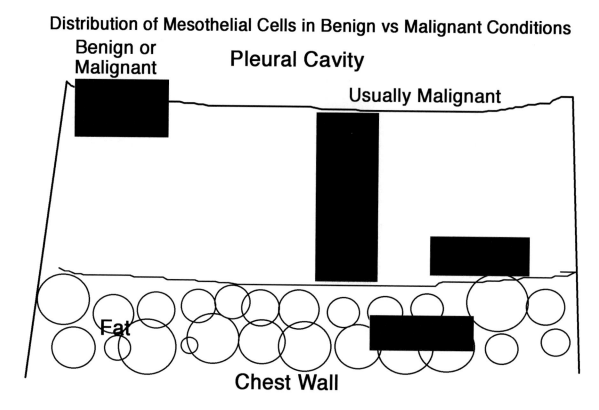

Figure 5-8

SCHEMATIC OF A SECTION OF A THICKENED PLEURA

The distribution of the mesothelial cells (in red) aids in the diagnosis of malignancy. Epithelial mesothelial proliferations that extend completely through the pleura are usually malignant, and those that enter the fat of the chest wall are almost always malignant. (Fig. 1 from Churg A, Colby TV, Cagle P, et al. The separation of benign and malignant mesothelial proliferations. Am J Surg Pathol 2000;24:1185.)

presence of invasive tumor, the lesion appeared to have been benign on follow-up (8).

At this point there appear to be no reliable morphologic criteria for the diagnosis of mesothelioma in situ, and we suggest that the term not be used. Such processes should instead be labeled "atypical mesothelial hyperplasia" or "atypical mesothelial proliferation" and accompanied by a request for further tissue.

These considerations are meant to apply to single layers of cells and masses of heaped up cells. An obvious bulk tumor spreading along the serosal surface should be diagnosed as malignant.

Mesothelial Proliferations Invading Underlying Tissue

The literature on the separation of benign and malignant mesothelial processes is scanty, but most recent publications emphasize the im-

portance of stromal invasion (1,4,8,12,18). True invasion of the stroma is the most reliable criterion of malignancy when dealing with mesothelial proliferations. Inflammatory reactions, however, tend to trap mesothelial cells in granulation tissue or even dense fibrous tissue, and thus determining what is true stromal invasion can be difficult. Such a determination is straightforward when dealing with a biopsy that is completely replaced by tumor or when there are obvious tumor nodules. But when the amount of putative tumor in a biopsy is small or the cells appear very bland, or the biopsy itself is very small, then proving that stromal invasion is present is often a problem.

The distribution of mesothelial cells in tissue is often quite useful in separating benign from malignant proliferations. Figure 5-8 shows an approach to this issue using a schematic cross

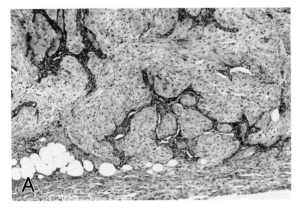

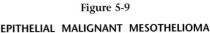

Figure 5-9

EPITHELIAL MALIGNANT MESOTHELIOMA

This lesion (A) could be mistaken for a benign process. But the lesion extends through the full thickness of the pleura into the fat (B), a sign of malignancy. Tumor cells appear deceptively innocuous, although nucleoli are present (C). Note the outward branching glands, a typical feature of epithelial mesotheliomas.

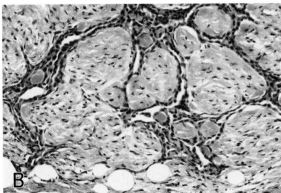

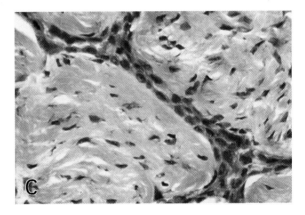

section of a thickened pleura. Mesothelial proliferations that invade only shallowly into a thickened pleura tend to be benign, although exceptions certainly exist, whereas proliferations that extend from the serosal surface to the junction with the chest wall fat, or that enter the fat, are usually malignant; indeed, in the pleural cavity, mesothelial proliferations that have invaded the fat of the chest wall should be regarded as malignant unless there is good reason to believe otherwise. Another way to view this is that the deeper the invasion, the more likely the process is to be malignant (figs. 5-9–5-11).

Occasionally, a biopsy samples a lesion in such a way that proliferating mesothelial cells are only found deep in a thickened pleura and not near the surface (fig. 5-10). This distribution is also strongly suggestive of malignancy, because benign organizing effusions tend to be more cellular and to trap mesothelial cells in the inflammatory reaction near the free pleural surface; such benign reactions usually become less and less cellular as one moves away from the pleural surface (see Spindle Cell Prolifera-

tions, below). Invasion of epithelial-type mesothelial cells into other structures, such as chest wall muscle or lung, is also a reliable indicator of malignancy.

Apparent cytologic blandness is not a reliable guide in this setting. Many epithelial mesotheliomas appear relatively monotonous (see chapter 4), and, at first glance, quite bland (figs. 5-9, 5-10, 5-12). Although careful examination shows that they usually have large nucleoli; these may be difficult to see when the epithelial lesion is composed of glands with flattened cells (fig. 5-9C), but should be searched for. In some unequivocally invasive tumors large nucleoli are hard to find. The presence or absence of mitoses is similarly of little help: epithelial mesotheliomas tend to have few mitoses, whereas mitoses are frequently a part of reactive mesothelial proliferations.

An additional helpful finding in separating apparently innocuous looking mesotheliomas from reactive mesothelial proliferations is the presence of nodular expansion of the stroma (figs. 5-11A; 5-12, left). Typically, the stroma in

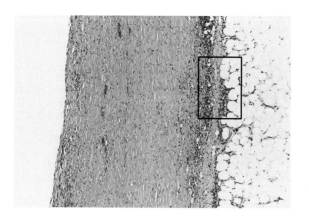
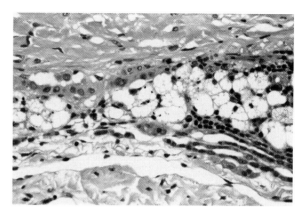

Figure 5-10

EPITHELIAL MALIGNANT MESOTHELIOMA

The lesion, in this biopsy, only is visible as glands in the deep portion of the pleura (box) and in the superficial fat. This distribution indicates that the process is malignant.

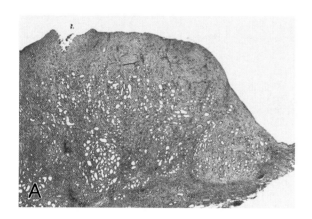
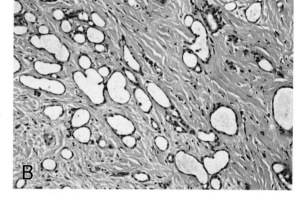

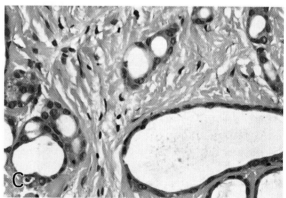

Figure 5-11

EPITHELIAL MALIGNANT MESOTHELIOMA

The full-thickness spread through the pleura and the small nodular stromal expansion in the lower right (A) are both signs of malignancy in a cytologically bland-appearing tumor (B,C). This lesion could be mistaken for a benign adenomatoid tumor, but such tumors are extremely rare in the pleura and do not form the type of stromal nodules with full-thickness spread seen here.

these regions is pale staining and sharply demarcated from the surrounding stroma. Nodular stromal expansion is a feature of mesotheliomas and not of reactive mesothelial processes.

These comments apply only to epithelial mesothelial cells. Infiltrating inflammatory cells may be found as a part of any mesothelial proliferation, but "invasion" by inflammatory cells has no particular significance. Occasionally, mesothelial reactions contain large numbers of eosinophils (fig. 5-13); this finding is not specific and does not indicate the presence

 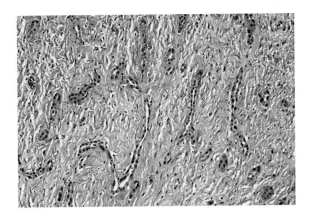

Figure 5-12

EPITHELIAL MALIGNANT MESOTHELIOMA

A distinct expansile stromal nodule and full-thickness spread through the pleura (left) are seen in a cytologically bland tumor (right).

 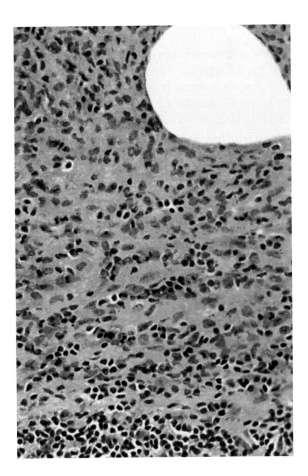

Figure 5-13

BENIGN REACTIVE PROCESS

Eosinophils are mixed with mesothelial cells.

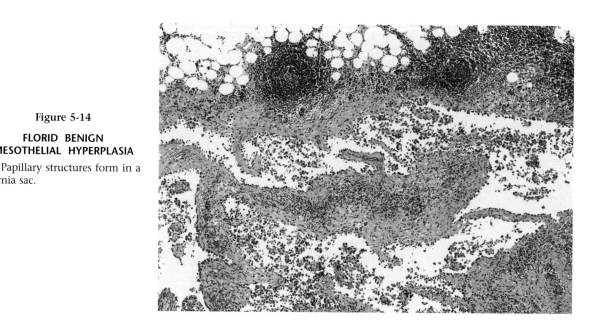

Figure 5-14

FLORID BENIGN MESOTHELIAL HYPERPLASIA

Papillary structures form in a hernia sac.

of any systemic process associated with eosinophilia. Reactive spindle-shaped mesothelial cells are an integral part of inflammatory reactions in the serosal membranes and can enter fat and even muscle. These cells also are not invasive, although they do have to be distinguished from sarcomatous mesotheliomas (see Spindle Cell Proliferations, below).

The same rules apply to the other serosal membranes; however, in the peritoneal cavity more caution needs to be exercised because complex inflammatory reactions, especially around the uterine adnexa, may trap mesothelial cells in unusual patterns (see below). But invasion of fat or the wall of an organ strongly favors malignancy in the peritoneal cavity as well.

Inguinal hernia sacs need separate mention because they are frequent sites of mesothelial reactions, possibly secondary to intermittent bowel incarceration. Reactive lesions range from simple hyperplasia (see fig. 5-2), to layers of atypical surface mesothelial cells, to papillary proliferations or small nodular aggregates of mesothelial cells in the lumen of the hernia sac (fig. 5-14) (4,16). Epithelial mesotheliomas, however, can occasionally arise in hernia sacs (7, 17). Epithelial mesotheliomas found in hernia sacs may be part of a more diffuse peritoneal mesothelioma and this should be noted in the surgical pathology report.

Separating Invasion from Entrapment

Entrapment of mesothelial cells by inflammatory processes is very common in all serosal membranes, and the distinction from invasion is crucial. As noted above, inflammatory processes tend to show a distinct pattern of organization (zonation), with more cellular areas near the free pleural/peritoneal and other serosal surfaces, and progressive loss of cellularity and increasing fibrosis as one moves away from the free surface. In the pleural cavity this process is often referred to as *organizing* or *fibrosing pleuritis/ pleurisy*. In organizing pleuritis, the mesothelial cells tend to be spindled (see below), but epithelial forms are sometimes seen, typically near the surface. Oddly enough, in organizing pleuritis it is common to find a more or less continuous layer of chronic inflammatory cells at the junction of the fibrotic pleura and the chest wall fat, even when the immediately overlying pleural tissue has few inflammatory cells.

As a general rule, mesothelial proliferations in the midst of an active inflammatory process should be regarded as benign unless there is definite morphologic evidence of malignancy away from the inflammatory focus. Inflammation may entrap mesothelial cells as individual cells or small glands. Such cells frequently have large nucleoli and cytologic atypia, and can appear quite alarming (fig. 5-15), as is true to a greater

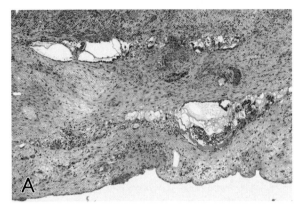

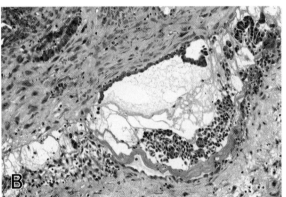

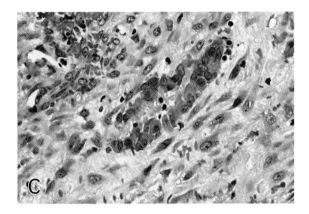

Figure 5-15

**ENTRAPPED MESOTHELIAL CELLS
IN AN INFLAMMATORY LESION
AROUND THE UTERINE ADNEXA**

Fibrin is mixed with the mesothelial cells (B), a good indication that these are entrapped. The high-power microscopic view (C) illustrates the marked atypia that may be seen in entrapped mesothelial cells.

or lesser extent of epithelial cells in the midst of inflammation in any area of the body. The combination of proliferating mesothelial cells, whether epithelial or spindled, and fibrin almost always represents entrapment, even if the focus appears to be deep to the free surface (fig. 5-15B).

A common feature of inflammatory mesothelial proliferations is the formation of linear arrays of individual mesothelial cells or small glands (fig. 5-16) (14). These arrays are derived from layering an inflammatory reaction on a preexisting mesothelial surface, with subsequent organization and entrapment of the mesothelial cells in the middle, or from the inflammatory adhesion of two preexisting mesothelial surfaces (fig. 5-16C). Again, the individual cells may be cytologically disturbing, but linear arrays are not a feature of malignant mesotheliomas, which tend to grow in a random pattern. Linear arrays are commonly seen in inflammatory processes around the uterine adnexa, but are also found in the pleural cavity and in hydroceles (fig. 5-17). Sometimes, multiple linear arrays are formed as inflammatory reac-

tions wax and wane and each recurrence deposits a new layer of fibrin which organizes on the old mesothelial surface.

When there is active ongoing inflammation the above-described foci are easy to diagnose as reactive, but the same type of linear processes may be seen in areas where the inflammation has receded and only fairly dense fibrous tissue remains. Because these foci have been formed by repeated layering, mesothelial cells may be present deep to the free surface. One helpful guideline is that these benign reactions, even if spread over a wide area, always seem to penetrate to exactly the same depth in the tissue. This phenomenon is particularly frequent, in our experience, in hydroceles (fig. 5-17A) and in the pleura, but occurs in the other serosal membranes as well. Keratin stains are useful in this context because they highlight the pattern of mesothelial cell growth and the sharp limit of downward spread of the mesothelial cells (fig. 5-17A).

In contrast, malignant mesotheliomas tend to invade without any special orientation, merely growing through preexisting tissue.

Figure 5-16

**ENTRAPPED MESOTHELIAL CELLS
FROM AN INFLAMMATORY MASS
AROUND THE UTERINE ADNEXA**

A,B: The cells are in linear arrays wih complicated papillary excrescences and pinched-off glands.

C: Low-power microscopic view. On the right side of the field, two mesothelial-lined surfaces are in apposition (arrow), illustrating how many entrapment lesions develop.

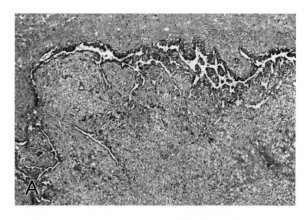

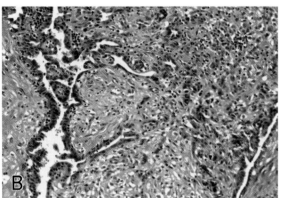

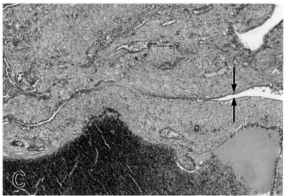

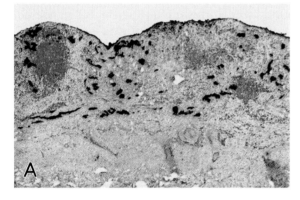

Figure 5-17

**ENTRAPPED MESOTHELIAL CELLS
IN A HYDROCELE SAC**

The keratin stain (A) shows that the proliferating cells, although numerous, only penetrate to a limited depth, i.e., do not extend all the way through the wall of the hydrocele. Hematoxylin and eosin (H&E) stain (B,C) shows the marked cytologic atypia that may be seen in entrapment lesions, even in old ones such as this where the inflammatory process has largely resolved.

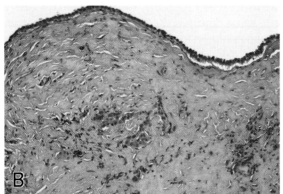

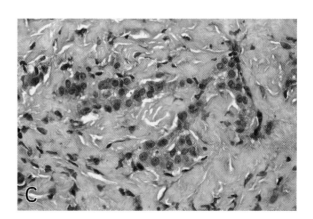

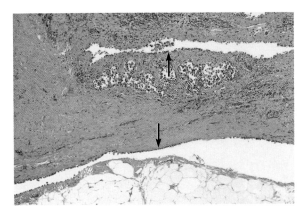 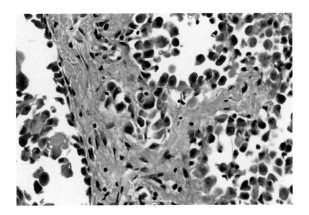

Figure 5-18

SPURIOUS APPEARANCE OF INVASION BY MESOTHELIAL CELLS

Low- (left) and high-power (right) views of mesothelial cells that line two surfaces (arrows), a certain indicator that most of the tissue is cut en face. These illustrations are from a hernia sac, but a similar appearance is common in small biopsies.

Invasion of fat by mesothelial cells in any location is almost always a sign of malignancy, and other criteria that apply to the pleura have been described above. Hydroceles can present a problem because benign reactions are, in our experience, often more florid in this location. Deep, invasive, irregular downgrowth and, particularly, invasion of muscle, are signs of malignancy in hydroceles (3).

All of the above comments imply that one has a large enough biopsy specimen that orientation and invasion are easily discerned. In small biopsy specimens containing only a small number of questionable cells, this may not be feasible because it is impossible to determine whether apparent invasion really represents en face cuts of the mesothelial surface (4,18). We advise proceeding with considerable caution in this circumstance; a diagnosis of atypical mesothelial proliferation with a request for further tissues is far preferable to an inaccurate diagnosis of mesothelioma based on equivocal invasion. One helpful guide in this circumstance is the presence of a mesothelial lining on two surfaces (fig. 5-18), an almost certain indicator that the section shows an en face cut.

Cytologic Atypia/Necrosis

As the above makes clear, cytologic atypia is generally not a useful guide to the separation of benign and malignant epithelial mesothelial proliferations. Not only may mesotheliomas be remarkably bland appearing, but reactive proliferations, particularly those at the surface/edge of an organizing effusion or those entrapped in an inflammatory reaction, are frequently quite atypical. Occasionally, epithelial mesotheliomas are cytologically very atypical (see chapter 4), but this is unusual, and in this circumstance one should consider whether the lesion is really a carcinoma, since carcinomas metastatic to the serosal membranes are generally much more anaplastic than epithelial mesotheliomas. Necrosis is uncommon in epithelial mesotheliomas, but when found, is strongly suggestive of malignancy. Necrosis may occasionally be seen in benign pleural reactions, particularly in tuberculosis, and after talc pleurodesis (fig. 5-19).

SPINDLE CELL PROLIFERATIONS

In the pleural cavity, spindle cell mesothelial proliferations present a diagnostic choice between organizing pleuritis and sarcomatous mesothelioma. Sarcomatous mesotheliomas are described and illustrated in chapter 4. In general, such tumors are densely cellular and the cells are packed in fascicles or storiform patterns typical of sarcomas; they rarely present a diagnostic problem. The real issue in this area is separating organizing pleuritis from desmoplastic mesothelioma, a problem that applies mostly to the pleural cavity; purely sarcomatous and desmoplastic mesotheliomas are rare in the peritoneal cavity and rare to nonexistent in hydroceles. A general approach is shown in Table 5-2.

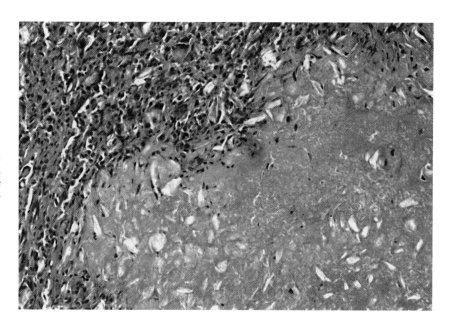

Figure 5-19

NECROSIS IN THE MIDST OF A MESOTHELIAL REACTION SECONDARY TO TALC PLEURODESIS

The talc appears as pale plates. Although necrosis is usually an indicator of malignancy in the pleura, it can, as here, occasionally be seen in benign reactions.

Table 5-2

SEPARATION OF DESMOPLASTIC MESOTHELIOMA FROM ORGANIZING PLEURITIS

Organizing Pleuritis[a]	Desmoplastic Mesothelioma[b]
Cellularity greatest immediately under effusion; becomes more fibrotic away from effusion (i.e., shows "zonation")	No zonation; bulk of lesion is paucicellular; may have abrupt transitions to cellular, frankly sarcomatous foci anywhere in the lesion
Cells immediately under effusion may be very atypical	Cytologic atypia often hard to discern
Capillaries perpendicular to pleural surface	Capillaries inconspicuous
No stromal invasion	Stromal invasion
No necrosis	Bland necrosis
No sarcomatous foci	Sarcomatous foci
No nodular expansion of stroma	Nodular expansions of stroma sometimes present

[a]Also called fibrous pleurisy and organizing pleurisy.

[b]Combination of paucicellular pattern plus stromal invasion, bland necrosis, or sarcomatous foci required for diagnosis.

Organizing Pleuritis

Organizing pleuritis is also referred to as *fibrous pleurisy, fibrosing pleuritis/pleurisy,* and *organizing pleurisy.* When the process is extreme and results in a dense fibrous rind surrounding and compressing the lung, it is referred to as a *fibrothorax.*

Organizing pleuritis reflects organization of a fibrinous and/or hemorrhagic pleural effusion (or recurrent effusions), or sometimes an empyema. In active lesions, fibrin is present on the pleural surface and often admixed with the superficial underlying cells. As the process ages, fibrin may totally disappear. The pleura is al-ways thickened in organizing pleuritis and usually shows a distinct zonation, with relatively cellular areas toward the free pleural surface and decreasing cellularity/increasing fibrous tissue as one moves away from the pleural surface (figs. 5-20–5-22). This zonation is extremely helpful in separating organizing pleuritis from sarcomatous mesotheliomas since the latter do not show any particular pattern of zonation and are often more cellular away from the pleural surface.

The cells at or just below the surface in organizing pleuritis may be very atypical when active inflammation is present. They are usually

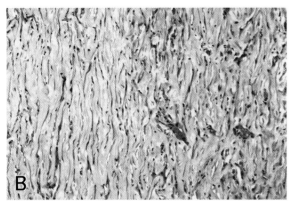

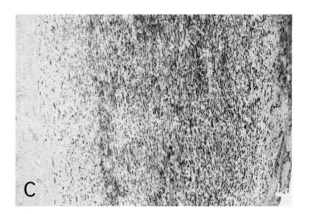

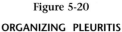

Figure 5-20

ORGANIZING PLEURITIS

A: Typical zonation is seen, with more cellular areas near the free pleural surface (right of field) and progressive increase in fibrous tissue toward the chest wall (left of field).

B: Higher-power view shows parallel layers of mesothelial cells in slit-like spaces. This process presumably represents layering of repeated effusions.

C: Keratin stain demonstrates both the layering and the loss of cells toward the chest wall. Reactive mesothelial processes such as this are almost always keratin positive.

spindle cells (fig. 5-23) but epithelial forms are sometimes seen (fig. 5-21, right). The cytologic atypia disappears as one moves deeper into the lesion and the cells become relatively inconspicuous and fewer in number, in many instances disappearing completely, to be replaced by fibrous tissue (figs. 5-20–5-23).

In most instances the fibrous tissue does not show any specific pattern, but storiform areas are sometimes present; by themselves, storiform foci are not diagnostic of desmoplastic mesothelioma, although very extensive storiform areas should bring desmoplastic mesothelioma to mind. The spindle cells may be very thin forms between bands of collagen or quite plump in more active lesions. In older lesions, the cells often appear to be in layers parallel to the surface (fig. 5-20B,C), a configuration that probably represents repeated layering of an ongoing or recurrent effusion. In active lesions the cells of organizing pleuritis are always positive with broad-spectrum keratin stains, and such stains are particularly useful to show the typical zonation and layering (fig. 5-20C). As the lesion ages,

keratin-positive cells may completely disappear, again a useful finding since desmoplastic mesotheliomas are, for all practical purposes, always keratin positive (4,10). But contrary to popular belief, keratin stains are not a marker of malignancy; they are usually positive in sarcomatous mesotheliomas but are commonly positive in benign reactions as well. Their utility lies in showing the distribution of mesothelial cells.

A process morphologically identical to organizing pleuritis occurs in hydroceles. There is zonation, with decreasing cellularity away from the hydrocele lumen, and, frequently, the same type of parallel layering seen in organizing pleuritis (3). As described above, linear arrays of entrapped mesothelial cells are common in hydroceles. A similar organizing pattern of zonation can be found in the peritoneal cavity, but is uncommon.

A variable infiltrate of chronic inflammatory cells is common in organizing pleuritis; like the mesothelial cells, this infiltrate tends to disappear deeper in the lesion. A feature that points to the benign nature of the lesion is the presence

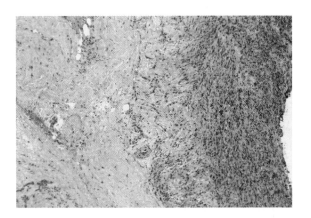
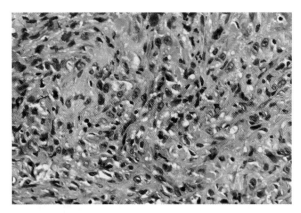

Figure 5-21

ORGANIZING PLEURITIS

Left: Marked cellularity is seen toward the free pleural surface.

Right: The higher-power view shows marked cytologic atypia. This is common at the surface of organizing pleuritis and is not an indicator of malignancy.

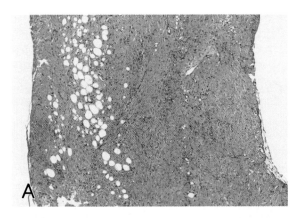

Figure 5-22

ORGANIZING PLEURITIS

Low-power (A) and high-power (B) microscopic views of organizing pleuritis. The lesion extends into the fat of the chest wall. There are no spindle cells between the fat cells (the usual finding in desmoplastic mesothelioma) and small vessels are present in the fibrous tissue, again not a feature of desmoplastic mesothelioma. Keratin stain (C) is negative, a helpful finding, since desmoplastic mesotheliomas are almost always keratin positive.

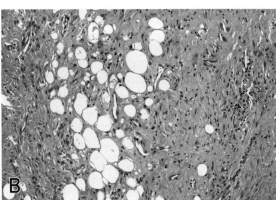
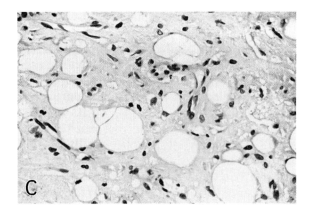

of elongated capillaries running perpendicular to the pleural surface (fig. 5-23) (4,10); these are not a feature of sarcomatous mesothelioma.

Organizing pleuritis often stops abruptly at the junction with the chest wall fat, and, as noted above, collections of lymphocytes may be present at this location. In some instances, however, the process extends into the fat (fig. 5-22), and, occasionally, into the chest wall muscle. Extension into the fat may mimic

 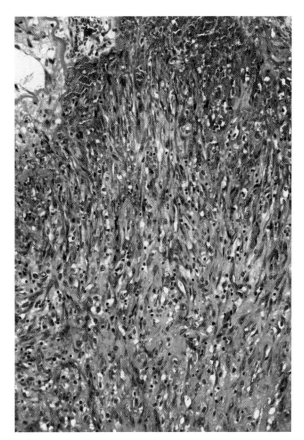

Figure 5-23

ORGANIZING PLEURITIS

A very active organizing pleuritis shows fibrin admixed with the cells at the free pleural surface. Perpendicular capillaries are prominent; these are a feature of benign organizing processes.

desmoplastic mesothelioma (see chapter 4), but the typical pattern of small fascicles of spindled cells separating fat cells that is characteristic of desmoplastic mesothelioma (4,10) is not seen in organizing pleuritis. Rather, the benign process is usually paucicellular and results in the production of fibrous tissue between the fat cells, with only occasional spindled cells. Small vessels are usually present in the fibrous tissue (fig. 5-22) and this is a helpful finding, since desmoplastic mesothelioma does not form vessels as it invades. Again, keratin stains may be of value in this setting because totally negative staining is evidence against a diagnosis of desmoplastic mesothelioma.

In general, organizing pleuritis reflects the presence of an underlying benign inflammatory process. Organizing pleuritis, however, can also be the result of a malignancy-induced pleural effusion. Thus, even when the diagnosis of organizing pleuritis appears obvious, the tissue should still be carefully examined for evidence of a neoplasm.

Fibrothorax represents an extreme example of organizing pleuritis in which a prior inflammatory process results in more or less total replacement of the fat by dense paucicellular fibrous tissue, often with extension of fibrous tissue into the chest wall muscle. Fibrothorax causes functional restriction, with marked contraction of the lung (fig. 5-24). These lesions are commonly calcified.

A proper biopsy is crucial to a correct diagnosis of organizing pleuritis. Needle biopsies are generally inadequate and even small thoracoscopic biopsies may not allow evaluation of zonation. If larger biopsies or pleural strippings

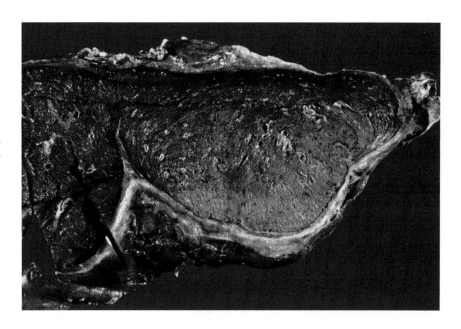

Figure 5-24

FIBROTHORAX

Dense fibrous tissue surrounds and contracts the underlying lung.

are provided, it is useful to put as much as possible of the tissue through for histologic examination, since the diagnostic areas of desmoplastic mesotheliomas are sometimes very localized.

Desmoplastic Mesothelioma

The features of desmoplastic mesothelioma are listed in Table 5-2 and described and illustrated in detail in chapter 4. The crucial features are a paucicellular, densely fibrotic pattern of mesothelial cell proliferation, usually in a storiform or "patternless pattern," combined with evidence of stromal invasion, bland necrosis, or overtly sarcomatous foci (4,10). There is some overlap in the paucicellular patterns of organizing pleuritis and desmoplastic mesothelioma, although organizing pleuritis generally does not have extensive areas of storiform pattern. The crucial distinction is that organizing pleuritis does not show necrosis, stromal invasion, or overtly sarcomatous foci. Occasionally in desmoplastic mesothelioma, distinct expansile stromal nodules are seen; these are not a feature of organizing pleuritis where the stromal reaction tends to be fairly homogeneous.

ROLE OF IMMUNOHISTOCHEMISTRY

This topic has been left for last for the simple reason that, at this point, there is no immunochemical stain that reliably identifies a lesion as benign or malignant. The fact that both types of lesion are broad-spectrum keratin positive has been noted above. It has been suggested that malignant mesotheliomas are more often positive for p53 and epithelial membrane antigen (EMA) than reactive processes (2,11,15,20). This is true in a statistical sense in some studies (2,11, 15) but not in others (10,15), and in fact, Roberts et al. (15) recently reported that they found staining for p53 and EMA in a majority of reactive mesothelial proliferations. Conversely, a proportion of mesotheliomas do not stain with EMA or p53 (10,15). Similarly, it has been claimed that desmin is a marker of reactive mesothelial but not malignant cells, at least in effusion cytology specimens (5), but Hurlimann (9) found that, in histologic sections, mesotheliomas stained for desmin in 56 percent of cases and Gonzalez-Lois et al. (6) reported positive staining in 47 percent of cases. In the individual case, immunochemical staining reactions are simply too variable to be relied upon (see fig. 5-4) and we do not recommend their use.

REFERENCES

1. Battifora H, McCaughey WT. Tumors of the serosal membranes. Atlas of Tumor Pathology, 3rd Series, Fascicle 15. Washington DC: Armed Forces Institute of Pathology; 1994.
2. Cagle PT, Brown RW, Lebovitz RM. p53 immunostaining in the differentiation of reactive processes from malignancy in pleural biopsy specimens. Hum Pathol 1994;25:443–8.
3. Churg A. Paratesticular mesothelial proliferations. Semin Diagn Pathol 2003;20:272–8.
4. Churg A, Colby TV, Cagle P, et al. The separation of benign and malignant mesothelial proliferations. Am J Surg Pathol 2000;24:1183–200.
5. Davidson B, Nielsen S, Christensen J, et al. The role of desmin and N-cadherin in effusion cytology: a comparative study using established markers of mesothelial and epithelial cells. Am J Surg Pathol 2001;25:1405–12.
6. Gonzalez-Lois C, Ballestin C, Sotelo MT, Lopez-Rios F, Garcia-Prats MD, Villena V. Combined use of novel epithelial (MOC-31) and mesothelial (HBME-1) immunohistochemical markers for optimal first line diagnostic distinction between mesothelioma and metastatic carcinoma in pleura. Histopathology 2001;38:528–34.
7. Grove A, Jensen ML, Donna A. Mesotheliomas of the tunica vaginalis testis and hernial sacs. Virchows Arch A Pathol Anat Histopathol 1989;415:283–92.
8. Henderson DW, Shilkin KB, Whitaker D. Reactive mesothelial hyperplasia vs mesothelioma, including mesothelioma in situ: a brief review. Am J Clin Pathol 1998;110:397–404.
9. Hurlimann J. Desmin and neural marker expression in mesothelial cells and mesotheliomas. Hum Pathol 1994;25:753–7.
10. Mangano WE, Cagle PT, Churg A, Vollmer RT, Roggli VL. The diagnosis of desmoplastic malignant mesothelioma and its distinction from fibrous pleurisy: a histologic and immunohistochemical analysis of 31 cases including p53 immunostaining. Am J Clin Pathol 1998;110:191–9.
11. Mayall FG, Goddard H, Gibbs AR. p53 immunostaining in the distinction between benign and malignant mesothelial proliferations using formalin-fixed paraffin sections. J Pathol 1992;168:377–81.
12. McCaughey WT, Al-Jabi M. Differentiation of serosal hyperplasia and neoplasia in biopsies. Pathol Ann 1986;21(Pt 1):271–93.
13. McCaughey WT, Kannerstein M, Churg J. Tumors and pseudotumors of the serous membranes. Atlas of Tumor Pathology, 2nd Series, Fascicle 20. Washington, DC: Armed Forces Institute of Pathology; 1985.
14. McFadden DE, Clement PB. Peritoneal inclusion cysts with mural mesothelial proliferation. A clinicopathological analysis of six cases. Am J Surg Pathol 1986;10:844–54.
15. Roberts F, Harper CM, Downie I, Burnett RA: Immunohistochemical analysis still has a limited role in the diagnosis of malignant mesothelioma. A study of thirteen antibodies. Am J Clin Pathol 2001;116:253–62.
16. Rosai J, Dehner LP. Nodular mesothelial hyperplasia in hernia sacs: a benign reactive condition simulating a neoplastic process. Cancer 1975;35:165–75.
17. Tang CK, Gray GF, Keuhnelian JG. Malignant peritoneal mesothelioma in an inguinal hernia sac. Cancer 1976;37:1887–90.
18. Tuder RM. Malignant disease of the pleura: a histopathological study with special emphasis on diagnostic criteria and differentiation from reactive mesothelium. Histopathology 1986;10:851–65.
19. Whitaker D, Henderson DW, Shilkin KB. The concept of mesothelioma in situ: implications for diagnosis and histogenesis. Semin Diagn Pathol 1992;9:151–61.
20. Wolanski KD, Whitaker D, Shilkin KB, Henderson DW. The use of epithelial membrane antigen and silver-stained nucleolar organizer regions testing in the differential diagnosis of mesothelioma from benign reactive mesothelioses. Cancer 1998;82:593–90.

6 LOCALIZED BENIGN TUMORS AND TUMOR-LIKE CONDITIONS OF THE SEROSAL MEMBRANES

SOLITARY FIBROUS TUMOR

Solitary fibrous tumor (SFT) is a localized, well-circumscribed, usually benign, mesenchymal tumor of uncertain cell origin (23). SFT is also referred to as *localized fibrous tumor* and *submesothelial fibroma*, but *solitary fibrous tumor* is the current preferred term. In the past, SFT had gone by numerous other names in the literature, including localized mesothelioma, fibrous mesothelioma, and benign mesothelioma. Earlier use of the term "mesothelioma" in the labeling of these tumors arose from suspicions that these tumors might be mesothelial in origin, a hypothesis that has never been proven and has largely fallen by the wayside, particularly since SFT is now known to occur in many organs in the body (18,48,69,74,112,114). We strongly suggest that the term mesothelioma not be used for these tumors, since the appellation mesothelioma is likely to be interpreted as a diffuse malignant mesothelioma. As well, some pathologists interpret the term "fibrous mesothelioma" as synonymous with "sarcomatous mesothelioma," again a reason to call these localized lesions SFT.

SFT is uncommon, but is the most frequent primary localized neoplasm of the pleura (2,19, 24,25,48,67–69,87,88,103). Men and women are equally affected, and the average age is usually in the 50s, with a range from childhood to old age. Many patients are asymptomatic, but they may present with chest pain or shortness of breath.

Up to 80 percent of SFTs arise from the visceral pleura, but they also occur entirely within the lung parenchyma (usually subpleurally) or are attached to the diaphragm, parietal pleura, mediastinum, pericardium, and peritoneum (24, 25,48,69,74,112,114). Rarely, benign SFTs are multiple, and sometimes they recur (108). SFTs are also known to occur in many other sites in the body, most of them unrelated to serosal membranes (32,69). The majority of SFTs are benign and most are cured by local resection. Even patients with massive benign lesions may do well with reexpansion of the compressed lung after surgery (70). Rare malignant variants occur and are discussed in chapter 7.

Radiologic studies show a sharply demarcated pleural tumor (figs. 6-1, 6-2). On pathologic

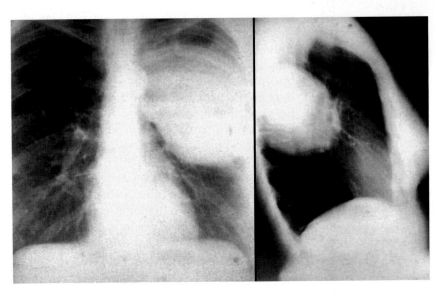

Figure 6-1

SOLITARY FIBROUS TUMOR

A large, circumscribed pleural mass is seen in the left upper chest (posteroanterior and lateral views).

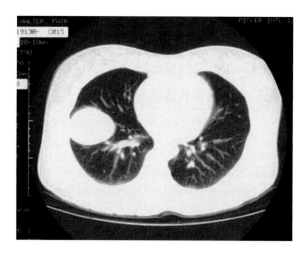

Figure 6-2

SOLITARY FIBROUS TUMOR

Computerized tomography (CT) discloses a sharply circumscribed, solid, pleural-based mass.

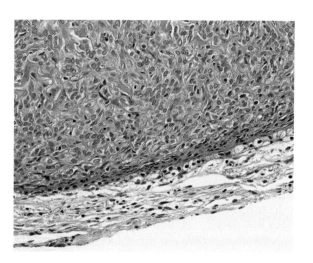

Figure 6-4

SOLITARY FIBROUS TUMOR

The border of the tumor is sharply demarcated from underlying tissue.

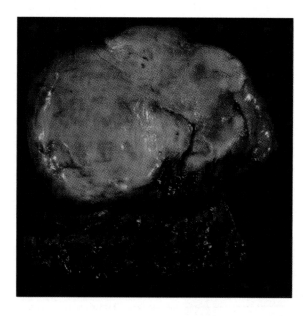

Figure 6-3

SOLITARY FIBROUS TUMOR

Lobulated, well-circumscribed mass with a glistening surface is attached to the visceral pleura.

examination, SFTs of the visceral pleura and other intrathoracic locations are typically well-circumscribed pleural-based nodules (fig. 6-3). They average 5 to 10 cm in diameter but sometimes grow to a diameter of 30 to 40 cm. These "giant" SFTs can fill the hemithorax, compress the adjacent lung, and produce symptoms, but

are slow growing and do not invade adjacent tissues unless they are malignant (70). About half of pleural-based SFTs are pedunculated and are attached to the visceral pleura by a stalk. As noted, most SFTs of the pleura do not cause symptoms, and when symptoms like chest pain or findings of pleural effusion are present, the likelihood of a malignant SFT increases, although a large benign SFT can also produce symptoms from lung compression. Both benign and malignant SFTs can cause hypoglycemia due to the production of insulin-like growth factor (*Doege-Potter syndrome*) (21,71,94,103).

Histologically, SFTs appear as sharply circumscribed, rounded masses that have pushing, but not invading, borders with adjacent lung parenchyma (fig. 6-4). The neoplastic cells are morphologically bland and resemble fibroblasts but appear more primitive and are embedded in a collagenous background stroma. To a lesser extent, primitive bland polygonal cells are also present. There are three classic histologic patterns: 1) the "patternless pattern of Stout" consists of bands of thick ropey collagen with slit-like spaces containing thin, inconspicuous neoplastic cells; 2) the hemangiopericytoma-like pattern has branching staghorn vessels; and 3) the cellular pattern consists of collections of benign spindle cells of varying cellular density, sometimes in a storiform arrangement (figs. 6-5–

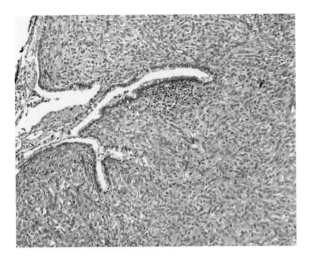

Figure 6-5

SOLITARY FIBROUS TUMOR

Airway epithelium entrapped within the solitary fibrous tumor should not be mistaken for an epithelial component.

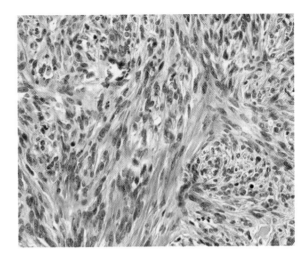

Figure 6-6

SOLITARY FIBROUS TUMOR

Spindle cells are arranged in fascicles.

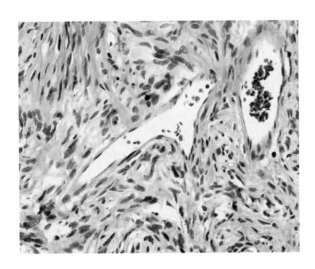

Figure 6-7

SOLITARY FIBROUS TUMOR

Leaf-like blood vessels resemble those seen in hemangiopericytoma.

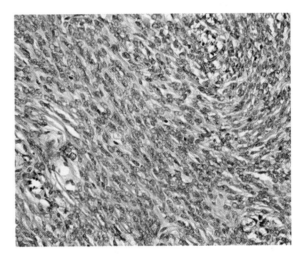

Figure 6-8

SOLITARY FIBROUS TUMOR

Cellular area of spindle and oval cells lacks the necrosis, pleomorphism, and high mitotic rate typical of malignancy.

6-9). Entrapped pulmonary epithelium may be present and gives a false impression that the SFT has an epithelial component (fig. 6-10). Focal stromal myxoid change is a common finding.

The neoplastic spindle cells of an SFT are characteristically immunopositive for CD34 and vimentin, and immunonegative for cytokeratins, smooth muscle actin, desmin, muscle-specific actin, S-100 protein, CD31, and factor VIII. This staining pattern assists in differentiating SFTs from other spindle cell neoplasms of the pleura including diffuse malignant mesothelioma (fig. 6-11) (2,12,39,48,68,88,109). Immunopositivity for bcl-2 protein has also been reported (50). The various chromosomal changes demonstrated by karyotypic analysis and comparative genomic hybridization do not have diagnostic significance at this time (73,82).

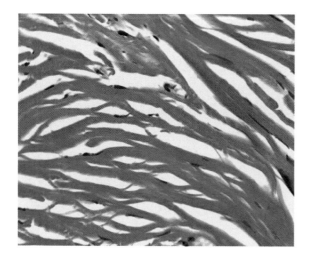

Figure 6-9

SOLITARY FIBROUS TUMOR

Less cellular areas consist of ropy bundles of collagen separated by slit-like spaces lined by flattened cells.

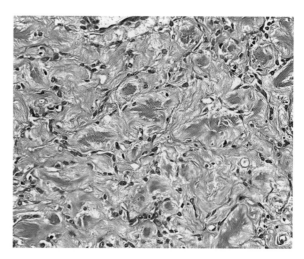

Figure 6-10

SOLITARY FIBROUS TUMOR

Collagen bundles are separated by slit-like spaces lined by cords of flattened cells.

Figure 6-11

SOLITARY FIBROUS TUMOR

Spindle cells are immunopositive for CD34 (A), immunopositive for vimentin (B), and immunonegative for keratin (C). Entrapped pulmonary epithelium in upper right corner of C is immunopositive for keratin.

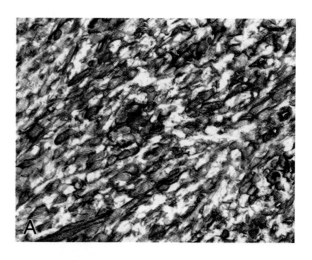

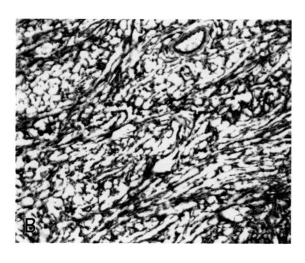

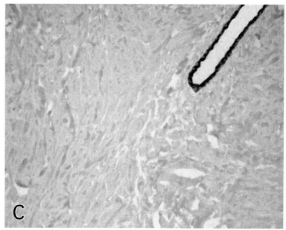

NODULAR PLEURAL PLAQUE

Pleural plaques are firm, circumscribed, ivory-colored collagenous structures on the pleural surface (24,25). There are multiple causes of pleural plaques, but classic parietal pleural plaques are caused by asbestos exposure. The plaques are on the surface of the lower portions of the parietal pleura or superior surface of the diaphragm and are often bilateral. While most plaques are flattened, pleural plaques can be nodular and give the radiographic impression of a pleural-based neoplasm (fig. 6-12).

Histologically, nodular pleural plaques show features of their less nodular counterparts (see chapter 8). They consist of dense, virtually acellular collagen in a "basket-weave" pattern (fig. 6-13). Calcifications and sometimes ossification may be present. They should be distinguished from desmoplastic mesotheliomas (see chapters 4 and 5), solitary fibrous tumors, and other fibrous lesions of the pleura.

WELL-DIFFERENTIATED PAPILLARY MESOTHELIOMA

Well-differentiated papillary mesothelioma (WDPM) is a distinctive tumor with a papillary architecture and a tendency toward superficial spread without invasion. These tumors are usually found in the peritoneal cavity of women, but also occur in the pleura, pericardium, and

tunica vaginalis testis (7,17,22,31,99). WDPMs usually occur as diffuse multifocal lesions, but may occur as solitary localized masses. Localized variants have been reported in the peritoneum, pleura, and pericardium (79,95,113).

WDPMs are most frequently an incidental finding at the time of surgery, but ascites or pleural effusion, abdominal or chest pain, or

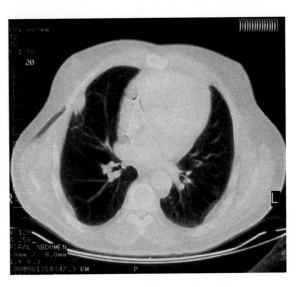

Figure 6-12

NODULAR PLEURAL PLAQUE

CT scan shows a circumscribed pleural-based mass.

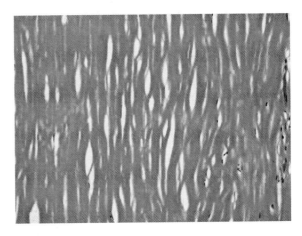

Figure 6-13

NODULAR PLEURAL PLAQUE

Left: Low-power microscopic view shows acellular collagen bundles in a "basket-weave" pattern characteristic of pleural plaque. Also present is underlying adipose tissue.

Right: Higher-power view shows only an occasional flattened nucleus lining slit-like spaces between collagen bundles.

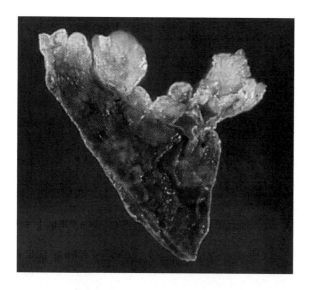

Figure 6-14

WELL-DIFFERENTIATED PAPILLARY MESOTHELIOMA

Papillary fronds arise from the visceral pleural surface of a wedge biopsy of lung tissue.

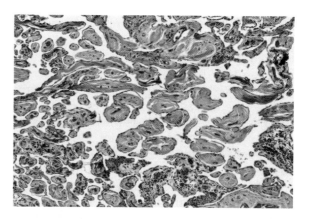

Figure 6-15

WELL-DIFFERENTIATED PAPILLARY MESOTHELIOMA

Single layers of bland flattened cells cover the surface of the papillary cores, which are composed of dense collagen.

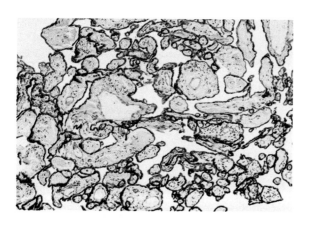

Figure 6-16

WELL-DIFFERENTIATED PAPILLARY MESOTHELIOMA

The flattened mesothelial cells that cover the surface of the papillary cores stain for keratin.

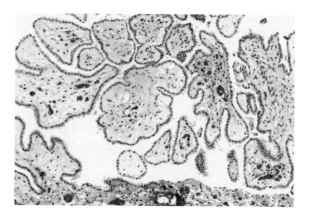

Figure 6-17

WELL-DIFFERENTIATED PAPILLARY MESOTHELIOMA

Single layers of bland cuboidal cells cover the surface of the papillary cores, which are composed of edematous myxoid stroma.

pneumothorax may be the presenting manifestation. Grossly, the tumor most often appears as multiple nodules ranging from a few millimeters to several centimeters in diameter; however, some cases have only a single nodule. The involved serosal surface may have a velvety appearance caused by the presence of numerous small tumor nodules (fig. 6-14). In the abdomen, small foci of tumor may be present on the ovarian or appendiceal surface.

Microscopically, WDPM is characterized by a broad fibrovascular core covered by a single layer of flattened to cuboidal mesothelial cells (figs. 6-15, 6-16). In some cases the core has a myxoid appearance (fig. 6-17) (33). Solid or tubular areas may also be observed. Psammoma bodies are occasionally seen and a central scar may be present, especially in the localized variant. The tumor cells are uniformly bland, and do not show mitotic figures, large nucleoli, or necrosis. Basal vacuoles may be present in the lining cells. Results of immunohistochemical staining are similar to those observed for the epithelial variant of diffuse malignant mesothelioma (see chapter 4).

The mesothelial nature of the lining cells has been confirmed ultrastructurally (17).

In the strictest definition, invasion is not observed in WDPM. Some otherwise typical cases, however, show limited invasion (17,23). It should be noted that diffuse malignant mesotheliomas may have areas with a WDPM-like pattern, and such cases should not be designated as WDPM (see chapter 4). Therefore, caution needs to be employed when attempting to make a diagnosis of WDPM on a small biopsy specimen. Although the cytologic features of WDPM have been described (46,51,60), cytologic specimens may fail to identify an invasive component that could have important prognostic implications. Cytologic diagnosis is not recommended.

Localized WDPM invariably pursues a benign course and complete surgical resection is curative. No association of localized WDPM and asbestos exposure has been reported. As a rule, tumors that consist of multiple small nodules are characterized by an indolent course with prolonged patient survival, and in most instances these tumors do not shorten life expectancy (16,17,31,53). In the original description by Daya and McCaughey (31), almost all tumors were multiple and the few patients with persisting but innocuous tumor who died, did so of other causes. More recent data suggest that on occasion these tumors are fatal (17,41). The development of invasive foci within WDPM may herald a more aggressive clinical course. Rapidly progressive disease suggests that the underlying process is actually a diffuse malignant mesothelioma with focal areas resembling WDPM, but that the diagnostic areas have not been biopsied. Asbestos exposure has been reported in some cases of multinodular WDPM (17,41), but an association has not been established in epidemiologic studies.

LIPOMA

Benign lipomatous tumors arising in subserosal tissues include *lipomas* and *lipoblastomas* of the pleura, mesentery, omentum, and epicardium and lipoblastomas of the parietal pleura, mesentery, and omentum (8,15,26,36, 43,44,57,61,63,64,75,76,89,97,101,105,116). Many of these occur in the pediatric age group and are rare. Lipomas arising in the chest wall

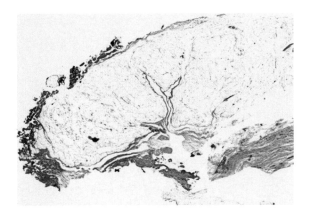

Figure 6-18

LIPOMA

Pleural lipoma consists of a well-circumscribed mass of mature adipose tissue. A thin fibrous capsule is attached to the underlying muscle and connective tissue of the pleura.

fat are not unusual. Occasionally, chest wall lipomas grow between the ribs to protrude into the pleural cavity. In a review of 4,000 computerized tomography (CT) scans of the chest, lipomas presenting as incidental pleural masses were identified in 6 cases (14,29). Although most subpleural lipomas in adults are asymptomatic incidental findings, one infarcted pedunculated intrathoracic lipoma produced chest pain and an inhomogeneous CT appearance that mimicked liposarcoma (42).

Histologically, lipomas consist of mature adipose tissue whereas lipoblastomas are composed of fetal adipose tissue (fig. 6-18). Surgical excision to establish the diagnosis usually removes the tumor and additional therapy is not needed for these benign lesions.

ADENOMATOID TUMOR

Adenomatoid tumors are benign mesothelial neoplasms that are usually incidental findings, primarily in the female and male genital tracts and rarely in extragenital locations such as the pleura and mesentery (fig. 6-19) (6,20,27,65,86, 115). Histologically, they consist of epithelioid cells that form vacuoles and tubular spaces within a fibrous stroma (fig. 6-20). The neoplastic cells are immunopositive for keratin and mesothelial markers, and immunonegative for vascular and carcinoma markers. Composite tumors consisting of adenomatoid tumor with

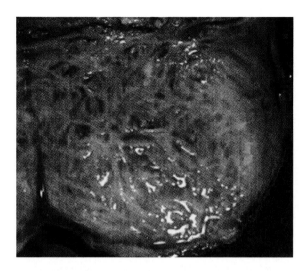

Figure 6-19

ADENOMATOID TUMOR

Cut surface of an adenomatoid tumor shows a well-circumscribed mass attached to the peritoneal surface of the uterus.

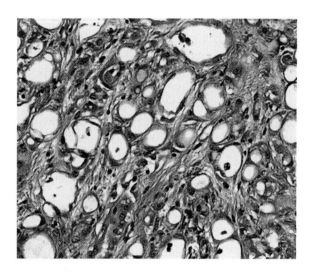

Figure 6-20

ADENOMATOID TUMOR

Mesothelial cells containing cytoplasmic vacuoles are present within a fibrous stroma.

multicystic mesothelioma of the peritoneum have been reported (20). Adenomatoid tumors are benign and no other treatment is indicated or required apart from surgical excision, which is usually done to obtain a diagnosis.

It is important to note that the few reported adenomatoid tumors of the pleura have been small circumscribed nodules. (65) Some forms of diffuse malignant mesothelioma can have a histologic appearance mimicking adenomatoid tumor (see chapter 4), but knowledge of the gross distribution of the tumor should make the distinction obvious.

CALCIFYING FIBROUS TUMOR

Calcifying fibrous tumor (CFT) is a rare benign fibrous lesion typically found in subcutaneous and deep soft tissues of the extremities, trunk, and neck of children, teenagers, and young adults (9,35,37,38,40,47,52,56,78,85,91–93,100, 110). A few cases have been reported in the pleura and mediastinum. Pinkard et al. (91) described three cases of CFT of the pleura in a 23-year-old woman, a 34-year-old man with chest pain, and a 28-year-old woman who was asymptomatic. A subsequent radiologic study showed that the pleural CFTs consisted of 3- to 12-cm pleural masses, all in the inferior chest; two patients had solitary masses and one had multifocal ipsilateral masses. The masses were

noncalcified on chest X ray but were found to be calcified on CT scan. Additional cases in the pleura, mediastinum, and chest wall have been reported, including miliary pleural lesions in a 29-year-old asymptomatic woman who also had a single soft tissue lesion (47).

Histologically, CFT consists of thick bundles of collagen, scattered fibroblasts, psammomatous or dystrophic calcifications, and a lymphoplasmacytic infiltrate (figs. 6-21, 6-22). The fibroblasts stain for vimentin and are negative for cytokeratins and CD34, features that distinguish CFT from diffuse malignant mesothelioma and solitary fibrous tumor, respectively. Some investigators have proposed that CFT is a late sclerosing stage of inflammatory myofibroblastic tumor, but others have highlighted histologic, immunohistochemical, and ultrastructural differences between these two entities. Surgical excision is generally all that is needed for treatment of CFT, but rare local recurrence has been reported.

SIMPLE MESOTHELIAL CYSTS OF THE PERITONEUM

Simple unilocular mesothelial cysts are generally thought to be hyperplastic reactive lesions or inclusion cysts rather than true neoplasms. They may or may not be related to the multicystic mesothelioma discussed below (34,77).

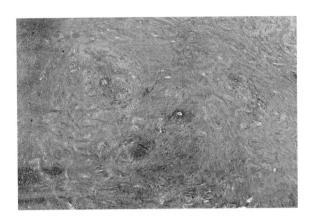

Figure 6-21

CALCIFYING FIBROUS TUMOR

Low-power microscopic view shows dense collagen and scattered inflammatory cells.

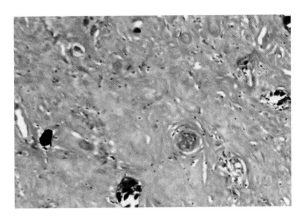

Figure 6-22

CALCIFYING FIBROUS TUMOR

Medium-power microscopic view of figure 6-21 shows dense collagen and calcifications.

Most simple mesothelial cysts are found in the peritoneum in women, and most are found incidentally; however, unilocular mesothelial cysts clinically mimicking renal masses and presenting with flank pain have been described.

Simple mesothelial cysts are fluid-filled and have thin translucent walls (fig. 6-23). They are generally no more than a few centimeters in greatest dimension. They may be multiple and can be free floating in the peritoneal cavity. Histologically, they consist of single layered, flattened to cuboidal mesothelial cells lining a thin fibrous wall (fig. 6-24). Positive immunostains for mesothelial markers confirm that the lining cells are mesothelial.

MULTICYSTIC MESOTHELIOMA OF THE PERITONEUM

Multicystic mesotheliomas are considered to be benign or indolent neoplasms by some authors and to be hyperplastic reactive lesions by others, some of whom prefer the appellation *multicystic* or *multilocular inclusion cysts* (1,3–5,10,11, 13,14,28–30,32,34,45,49,54,55,58,59,62,66,72, 74,77,80,81,83,84,90,96,98,104,106,107,111). As a compromise, the term *benign multicystic mesothelial proliferation* has been proposed.

Multicystic mesotheliomas occur most frequently in young to middle-aged premenopausal women and are found most often in the pelvic peritoneum, growing along the surfaces of the cul de sac, uterus, and rectum. Patients

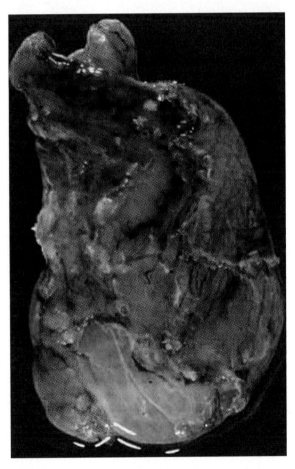

Figure 6-23

SIMPLE MESOTHELIAL CYST OF THE PERITONEUM

Thin, translucent sack on the surface of the omentum.

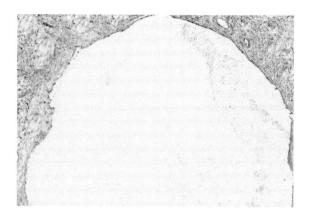

Figure 6-24

SIMPLE MESOTHELIAL CYST OF THE PERITONEUM

The cyst is lined by flattened mesothelial cells.

Figure 6-25

MULTICYSTIC MESOTHELIOMA OF THE PERITONEUM

Multiple, bubble-like, thin-walled cysts of peritoneal multicystic mesothelioma. (Courtesy of Dr. C. Ceballos, Vancouver, BC, Canada.)

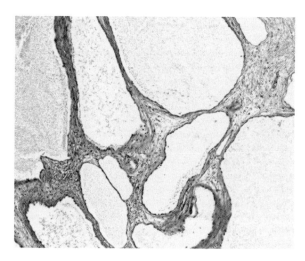

Figure 6-26

MULTICYSTIC MESOTHELIOMA OF THE PERITONEUM

Multiple cysts with thin fibrous walls are lined by flattened mesothelial cells. (Courtesy of Dr. A. Laga, Durham, NC.)

present with abdominal pain and, less often, abdominal swelling or mass. A few cases in men and rare cases in other locations, including the pleura, pericardium, and spermatic cord, have been reported. Some authors report an association of multicystic mesothelioma with previous abdominal surgery and endometriosis, which they consider support for its origin as a non-neoplastic reactive lesion. Other authors report the absence of previous inflammatory conditions, proposing instead that this observation supports a true neoplastic origin.

Many of these lesions are benign, but reports of recurrence and "transition" to diffuse malig-

nant mesothelioma may indicate that various authors have included more than one entity in this category or, alternatively, that there is a true spectrum of this disease.

Grossly, multicystic mesotheliomas consist of multiple or multilocular, fluid-filled cysts that have thin translucent walls (fig. 6-25). The cysts may spread along the serosal surface, form a multilocular mass, or be associated with additional unilocular cysts, including free-floating cysts. The cystic fluid is clear, serous, or gelatinous. Cytology of the cystic fluid shows sheets of benign monomorphous mesothelial cells.

Histologically, multicystic mesotheliomas are composed of cysts of varying size that are lined by flattened to cuboidal mesothelial cells with a delicate fibrovascular stroma (figs. 6-26, 6-27). Occasionally, the lining cells have a "picket-fence" or "hob-nail" appearance. Adenomatoid change or squamous metaplasia of the lining cells occurs in up to one third of cases. Positive immunostains for mesothelial markers and ultrastructural findings confirm that the cells lining the cysts are mesothelial.

SCHWANNOMA

Intrathoracic schwannomas usually occur in the paraspinal region of the mediastinum where

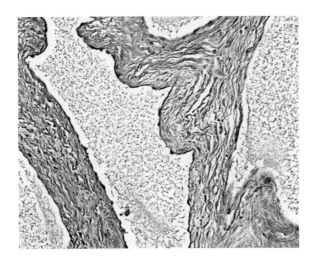

Figure 6-27

MULTICYSTIC MESOTHELIOMA OF THE PERITONEUM

High-power microscopic view shows flattened mesothelial cells lining the fibrous walls of multiple cysts. (Courtesy of Dr. A. Laga, Durham, NC.)

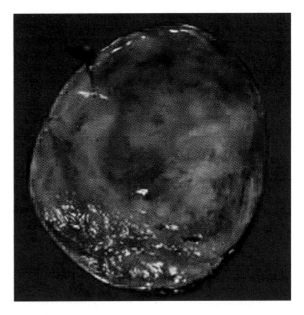

Figure 6-28

SCHWANNOMA

The cut surface of a well-circumscribed mass shows firm, tan-yellow tissue containing a small cyst.

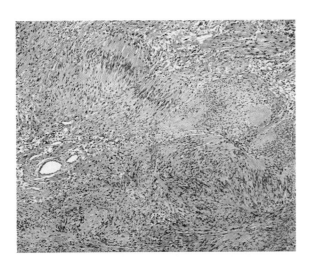

Figure 6-29

SCHWANNOMA

Palisading spindle cells are seen at low-power microscopy.

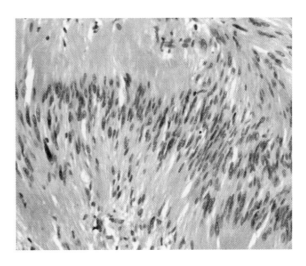

Figure 6-30

SCHWANNOMA

High-power microscopic view of figure 6-29 shows palisading spindle cells with parallel nuclei.

they may raise the possibility of a pleural tumor. Most intrathoracic schwannomas have the same general histopathologic features as schwannomas in other locations and are immunopositive for neural markers (figs. 6-28– 6-30). A rare form of schwannoma, the *melanotic schwannoma*, has cytoplasmic brown pigment and most commonly occurs in the paraspinal region where it involves spinal nerve roots and sympathetic ganglia.

REFERENCES

1. Adolph AJ, Smith TE, Adolph J. Benign multicystic mesothelioma: a case report. J Obstet Gynaecol Can 2002;24:246–7.

2. Ali SZ, Hoon V, Hoda S, Heelan R, Zakowski MF. Solitary fibrous tumor. A cytologic-histologic study with clinical, radiologic, and immunohisto-chemical correlations. Cancer 1997;81:116–21.

3. Alvarez-Fernandez E, Rabano A, Barros-Malvar JL, Sanabia-Valdez J. Multicystic peritoneal mesothelioma: a case report. Histopathology 1989; 14:199–208.

4. Baddoura FK, Varma VA. Cytologic findings in multicystic peritoneal mesothelioma. Acta Cytol 1990;34:524–8.

5. Ball NJ, Urbanski SJ, Green FH, Kieser T. Pleural multicystic mesothelial proliferation. The so-called multicystic mesothelioma. Am J Surg Pathol 1990;14:375–8.

6. Banks ER, Mills SE. Histiocytoid (epithelioid) hemangioma of the testis. The so-called vascular variant of "adenomatoid tumor." Am J Surg Pathol 1990;14:584–9.

7. Barbera V, Rubino M. Papillary mesothelioma of the tunica vaginalis. Cancer 1957;10:183–9.

8. Beebe MM, Smith MD. Omental lipoblastoma. J Pediatr Surg 1993;28:1626–7.

9. Ben-Izhak O, Itin L, Feuchtwanger Z, Lifschitz-Mercer B, Czernobilsky B. Calcifying fibrous pseudotumor of mesentery presenting with acute peritonitis: case report with immunohisto-chemical study and review of literature. Int J Surg Pathol 2001;9:249–53.

10. Benson RC Jr, Williams TH. Peritoneal cystic mesothelioma: successful treatment of a difficult disease. J Urol 1990;143:347–8.

11. Blumberg NA, Murray JF. Multicystic peritoneal mesothelioma. A case report. S Afr Med J 1981;59:85–6.

12. Bongiovanni M, Viberti L, Pecchioni C, et al. Steroid hormone receptor in pleural solitary fibrous tumours and CD34+ progenitor stromal cells. J Pathol 2002;198:252–7.

13. Bos SD, Jansen W, Ypma AF. Multicystic mesothelioma presenting as a pelvic tumour: case report and literature review. Scand J Urol Nephrol 1995;29:225–8.

14. Bui-Mansfield LT, Kim-Ahn G, O'Bryant LK. Multicystic mesothelioma of the peritoneum. AJR Am J Roentgenol 2002;178:402.

15. Buirski G, Goddard P. The C.T. diagnosis of pleural lipoma. Bristol Med Chir J 1986;101:33, 43.

16. Burrig KF, Pfitzer P, Hort W. Well-differentiated papillary mesothelioma of the peritoneum: a borderline mesothelioma. Report of two cases and review of the literature. Virch Arch A Pathol Anat Histopathol 1990;417:443–7.

17. Butnor KJ, Sporn TA, Hammar SP, Roggli VL. Well-differentiated papillary mesothelioma. Am J Surg Pathol 2001;25:1304–9.

18. Cagle, PT. Tumors of the lung (excluding lymphoid tumors). In: Thurlbeck WM, Churg AM, eds. Pathology of the lung, 2nd ed. New York: Thieme; 1995:437–552.

19. Cardillo G, Facciolo F, Cavazzana AO, Capece G, Gasparri R, Martelli M. Localized (solitary) fibrous tumors of the pleura: and analysis of 55 patients. Ann Thorac Surg 2000;70:1808–12.

20. Chan JK, Fong MH. Composite multicystic mesothelioma and adenomatoid tumour of the uterus: different morphological manifestations of the same process? Histopathology 1996;29:375–7.

21. Chaugle H, Parchment C, Grotte GJ, Keenan DJ. Hypoglycaemia associated with a solitary fibrous tumour of the pleura. Eur J Cardiothorac Surg 1999;15:84–6.

22. Chetty R. Well-differentiated (benign) papillary mesothelioma of the tunica vaginalis. J Clin Pathol 1992;45:1029–30.

23. Churg A, Roggli VL, Galateau-Salle F, et al. Tumours of the pleura. In: Travis WD, World Health Organization, Brambilla ER, et al, eds. Pathology and genetics of tumours of the lung, pleura, thymus and heart. Lyon: IARC Press; 2004:125–44.

24. Churg AM. Diseases of the pleura. In: Thurbeck WM, Churg AM, eds. Pathology of the lung, 2nd ed. New York: Thieme; 1995:1067–110.

25. Churg AM. Localized pleural tumors. In: Cagle PT, ed. Diagnostic pulmonary pathology. New York: Marcel Dekker; 2000:719–35.

26. Corbi P, Boufi M, Thierry G, Menu P. Giant pleural lipoma. Eur J Cardiothorac Surg 1999;16:249–50.

27. Craig JR, Hart WR. Extragenital adenomatoid tumor: evidence for the mesothelial theory of origin. Cancer 1979;43:1678–81.

28. Cregan P, Wiley S, Lee CH. Multicystic peritoneal mesothelioma in a 29 year old male. Aust N Z J Surg 1987;57:271–3.

29. Cusatelli P, Altavilla G, Marchetti M. Benign cystic mesothelioma of peritoneum: a case report. Eur J Gynaecol Oncol 1997;18:124–6.

30. Datta RV, Paty PB. Cystic mesothelioma of the peritoneum. Eur J Surg Oncol 1997;23:461–2.

31. Daya D, McCaughey WT. Well-differentiated papillary mesothelioma of the peritoneum. A clinicopathologic study of 22 cases. Cancer 1990;65:292–6.

32. Devaney K, Kragel PJ, Devaney EJ. Fine-needle aspiration cytology of multicystic mesothelioma. Diagn Cytopathol 1992;8:68–72.

33. Diaz LK, Okonkwo A, Solans EP, Bedrossian C, Rao MS. Extensive myxoid change in well-differentiated papillary mesothelioma of the pelvic peritoneum. Ann Diagn Pathol 2002;6:164–7.

34. Drut R, Quijano G. Multilocular mesothelial inclusion cysts (so-called benign multicystic mesothelioma) of pericardium. Histopathology 1999;34:472–4.

35. Dumont P, de Muret A, Skrobala D, Robin P, Toumieux B. Calcifying fibrous pseudotumor of the mediastinum. Ann Thorac Surg 1997;63:543–4.

36. Epler GR, McLoud TC, Munn CS, Colby TV. Pleural lipoma. Diagnosis by computed tomography. Chest 1986;90:265–8.

37. Erasmus JJ, McAdams HP, Patz EF Jr, Murray JG, Pinkard NB. Calcifying fibrous pseudotumor of pleura: radiologic features in three cases. J Comput Assist Tomogr 1996;20:763–5.

38. Fetsch JF, Montgomery EA, Meis JM. Calcifying fibrous pseudotumor. Am J Surg Pathol 1993;17:502–8.

39. Flint A, Weiss SW. CD-34 and keratin expression distinguishes solitary fibrous tumor (fibrous mesothelioma) of pleura from desmoplastic mesothelioma. Hum Pathol 1995;26:428–31.

40. Fukunaga M, Kikuchi Y, Endo Y, Ushigome S. Calcifying fibrous pseudotumor. Pathol Int 1997;47:60–3.

41. Galateau-Salle F, Vignaud JM, Burke L, et al. Well-differentiated papillary mesothelioma of the pleura: a series of 24 cases. Am J Surg Path 2004;28:534–40.

42. Geis JR, Russ PD, Adcock KA. Computed tomography of a symptomatic infracted thoracic lipoma. J Comput Tomogr 1988;12:54–6.

43. Giubilei D, Cicia S, Nardis P, Patane E, Villani RM. Lipoma of the omentum in a child. Radiology 1980;137:357–8.

44. Gonzalez-Crussi F, Sotelo-Avila C, deMello DE. Primary peritoneal, omental, and mesenteric tumors in childhood. Semin Diagn Pathol 1986;3:122–37.

45. Groisman GM, Kerner H. Multicystic mesothelioma with endometriosis. Acta Obstet Gynecol Scand 1992;71:642–4.

46. Haba T, Wakasa K, Sasaki M. Well-differentiated papillary mesothelioma of the pelvic cavity. A case report. Acta Cytolog 2003;47:88–92.

47. Hainaut P, Lesage V, Weynand B, Coche E, Noirhomme P. Calcifying fibrous pseudotumor (CFPT): a patient presenting with multiple pleural lesions. Acta Clin Belg 1999;54:162–4.

48. Hanau CA, Miettinen M. Solitary fibrous tumor: histological and immunohistochemical spectrum of benign and malignant variants presenting at different sites. Hum Pathol 1995;26:440–9.

49. Hasan AK, Sinclair DJ. Case report: calcification in benign cystic peritoneal mesothelioma. Clin Radiol 1993;48:66–7.

50. Hasegawa T, Matsuno Y, Shimoda T, Hirohashi S, Hirose T, Sano T. Frequent expression of bcl-2 protein in solitary fibrous tumors. Jpn J Clin Oncol 1998;28:86–91.

51. Hejmadi R, Ganesan R, Kamal NG. Malignant transformation of a well-differentiated peritoneal papillary mesothelioma. Acta Cytolog 2003;47:517–8.

52. Hill KA, Gonzalez-Crussi F, Chou PM. Calcifying fibrous pseudotumor versus inflammatory myofibroblastic tumor: a histological and immunohistochemical comparison. Mod Pathol 2001;14:784–90.

53. Hoekman K, Tognon G, Risse EK, Bloemsma CA, Vermorken JB. Well-differentiated papillary mesothelioma of the peritoneum: a separate entity. Eur J Cancer 1996;32A:255–8.

54. Hove Kanstrup M, Joergensen A, Grove A. Benign multicystic peritoneal mesothelioma. Acta Obstet Gynecol Scand 2002;81:1083–5.

55. Hutchinson R, Sokhi GS. Multicystic peritoneal mesothelioma: not a benign condition. Eur J Surg 1992;158:451–3.

56. Jeong HS, Lee GK, Sung R, Ahn JH, Song HG. Calcifying fibrous pseudotumor of mediastinum—a case report. J Korean Med Sci 1997;12:58–62.

57. Ilhan H, Tokar B, Isiksoy S, Koku N, Pasaoglu O. Giant mesenteric lipoma. J Pediatr Surg 1999;34:639–40.

58. Inman DS, Lambert AW, Wilkins DC. Multicystic peritoneal inclusion cysts: the use of CT guided drainage for symptom control. Ann R Coll Surg Engl 2000;82:196–7.

59. Iverson OH, Hovig T, Brandtzaeg P. Peritoneal, benign, cystic mesothelioma with free-floating cysts, re-examined by new methods. A case report. APMIS 1988;96:123–7.

60. Jayaram G, Ashok S. Fine needle aspiration cytology of well-differentiated papillary mesothelioma. Report of a case. Acta Cytolog 1988;32:563–6.

61. Jimenez JF. Lipoblastoma in infancy and childhood. J Surg Oncol 1986;32:238–44.

62. Kampschoer PH, Ubachs HM, Theunissen PH. Benign abdominal multicystic mesothelioma. Acta Obstet Gynecol Scand 1992;71:555–7.

63. Kaniklides C, Frykberg T, Lundkvist K. Paediatric mesenteric lipoma, an unusual cause of repeated abdominal pain. A case report. Acta Radiol 1998;39:695–7.

64. Kanu A, Oermann CM, Malicki D, Wagner M, Langston C. Pulmonary lipoblastoma in an 18-month-old child: a unique tumor in children. Pediatr Pulmonol 2002;34:150–4.

65. Kaplan MA, Tazelaar HD, Hayashi T, Schroer KR, Travis WD. Adenomatoid tumors of the pleura. Am J Surg Pathol 1996;20:1219–23.

66. Katsube Y, Mukai K, Silverberg SG. Cystic mesothelioma of the peritoneum: a report of five cases and review of the literature. Cancer 1982;50:1615–22.

67. Kawai T, Yakumaru K, Mikata A, Kageyama K, Torikata C, Shimosato Y. Solitary (localized) pleural mesothelioma. A light- and electron-microscopic study. Am J Surg Pathol 1978;2:365–75.

68. Keating S, Simon GT, Alexopoulou I, Kay JM. Solitary fibrous tumour of the pleura: an ultrastructural and immunohistochemical study. Thorax 1987;42:976–9.

69. Khalifa MA, Montgomery EA, Azumi N, et al. Solitary fibrous tumors: a series of lesions, some in unusual sites. South Med J 1997;90:793–9.

70. Khan JH, Rahman SB, Clary-Macy C, et al. Giant solitary fibrous tumor of the pleura. Ann Thorac Surg 1998;65:1461–4.

71. Kim JH, Kim JO, Kim SY, Na MH, Lim SP, Kim JM. Two cases of large solitary fibrous tumors of the pleura associated with fasting hypoglycemia. Eur Radiol 2001;11:819–24.

72. Kjellevold K, Nesland JM, Holm R, Johannessen JV. Multicystic peritoneal mesothelioma. Pathol Res Pract 1986;181:767–73.

73. Krismann M, Adams H, Jaworska M, Muller KM, Johnen G. Patterns of chromosomal imbalances in benign solitary fibrous tumors of the pleura. Virchows Arch 2000;437:248–55.

74. Kubota Y, Kawai N, Tozawa K, Hayashi Y, Sasaki S, Kohri K. Solitary fibrous tumor of the peritoneum found in the prevesical space. Urol Int 2000;65:53–6

75. Kwak JY, Ha DH, Kim YA, Shim JY. Lipoblastoma of a parietal pleura in a 7-month-old infant. J Comput Assist Tomogr 1999;23:952–4.

76. Levin DC, Matthay RA. Case report. Subpleural lipoma. Clin Notes Respir Dis 1975;14:15–6.

77. Lamovec J, Sinkovec J. Multilocular peritoneal inclusion cyst (multicystic mesothelioma) with hyaline globules. Histopathology 1996;28:466–9.

78. Maeda A, Kawabata K, Kusuzaki K. Rapid recurrence of calcifying fibrous pseudotumor (a case report). Anticancer Res 2002;22:1795–7.

79. McCaughey WT, Kannerstein M, Churg J. Tumors and pseudotumors of the serous membranes. Atlas of Tumor Pathology, 2nd Series, Fascicle 20. Washington, DC: Armed Forces Institute of Pathology; 1985:76–8.

80. McCullagh M, Keen C, Dykes E. Cystic mesothelioma of the peritoneum: a rare cause of 'ascites' in children. J Pediatr Surg 1994;29:1205–7.

81. Mennemeyer R, Smith M. Multicystic, peritoneal mesothelioma: a report with electron microscopy of a case mimicking intra-abdominal cystic hygroma (lymphangioma). Cancer 1979; 44:692–8.

82. Miettinen MM, el-Rifai W, Sarlomo-Rikala M, Andersson LC, Knuutila S. Tumor size-related DNA copy number changes occur in solitary tumors but not in hemangiopericytomas. Mod Pathol 1997;10:1194–2000.

83. Moghe GM, Krishnamurthy SC. Multicystic mesothelioma of the peritoneum. Indian J Gastroenterol 2001;20:202–3.

84. Moore JH Jr, Crum CP, Chandler JG, Feldman PS. Benign cystic mesothelioma. Cancer 1980;45:2395–9.

85. Nascimento AF, Ruiz R, Hornick JL, Fletcher CD. Calcifying fibrous 'pseudotumor': clinicopathologic study of 15 cases and analysis of its relationship to inflammatory myofibroblastic tumor. Int J Surg Pathol 2002;10:189–96.

86. Natarajan S, Luthringer DJ, Fishbein MC. Adenomatoid tumor of the heart: report of a case. Am J Surg Pathol 1997;21:1378–80.

87. Norton SA, Clark SC, Sheehan AL, Ibrahim NB, Jeyasingham K. Solitary fibrous tumour of the diaphragm. J Cardiovasc Surg 1997;38:685–6.

88. Ordonez NG. Localized (solitary) fibrous tumor of the pleura. Adv Anat Pathol 2000;7:327–40.

89. Park CH, Kim KI, Lim YT, Chung SW, Lee CH. Ruptured giant intrathoracic lipoblastoma in a 4-month-old infant: CT and MR findings. Pediatr Radiol 2000;30:38–40.

90. Pelosi G, Zannoni M, Caprioli F, et al. Benign multicystic mesothelial proliferation of the peritoneum: immunohistochemical and electron microscopical study of a case and review of the literature. Histol Histopathol 1991;6:575–83.

91. Pinkard NB, Wilson RW, Lawless N, et al. Calcifying fibrous pseudotumor of pleura. A report of three cases of a newly described entity involving the pleura. Am J Clin Pathol 1996;105: 189–94.

92. Reed MK, Margraf LR, Nikaidoh H, Cleveland DC. Calcifying fibrous pseudotumor of the chest wall. Ann Thorac Surg 1996;62:873–4.

93. Rosenthal NS, Abdul-Karim FW. Childhood fibrous tumor with psammoma bodies. Clinicopathologic features in two cases. Arch Pathol Lab Med 1988;112:798–800.

94. Roy TM, Burns MV, Overly DJ, Curd BT. Solitary fibrous tumor of the pleura with hypoglycemia: the Doege-Potter syndrome. J Ky Med Assoc 1992;90:557–60.

95. Sane AC, Roggli VL. Curative resection of a well-differentiated papillary mesothelioma of the pericardium. Arch Pathol Lab Med 1995;119:266–7.

96. Santucci M, Biancalani M, Dini S. Multicystic peritoneal mesothelioma. A fine structure study with special reference to the spectrum of phenotypic differentiation exhibited by mesothelial cells. J Submicrosc Cytol Pathol 1989;21:749–64.

97. Schulman H, Barki Y, Hertzanu Y. Case report: mesenteric lipoblastoma. Clin Radiol 1992;46:57–8.

98. Scucchi L, Mingazzini P, Di Stefano D, Falchi M, Camilli A, Vecchione A. Two cases of "multicystic peritoneal mesothelioma": description and critical review of the literature. Anticancer Res 1994;14:715–20.

99. Shukunami K, Hirabuki S, Kaneshima M, Kamitani N, Kotsuji F. Well-differentiated papillary mesothelioma involving the peritoneal and pleural cavities: successful treatment by local and systemic administration of carboplatin. Tumori 2000;86:419–21.

100. Sigel JE, Smith TA, Reith JD, Goldblum JR. Immunohistochemical analysis of anaplastic lymphoma kinase expression in deep soft tissue calcifying fibrous pseudotumor: evidence of a late sclerosing stage of inflammatory myofibroblastic tumor? Ann Diagn Pathol 2001;5:10–4.

101. Signer RD, Bregman D, Klausner S. Giant lipoma of the mesentery: report of an unusual case and review of the literature. Am Surg 1976;42:595–7.

102. Stocker JT, Dehnner LP. Acquired pediatric and neonatal diseases. In: Dail DH, Hammar SP, eds. Pulmonary pathology, 2nd ed. New York: Springer-Verlag; 1994:191–254.

103. Strom EH, Skjorten F, Aarseth LB, Haug E. Solitary fibrous tumor of the pleura. An immunohistochemical, electron microscopic and tissue culture study of a tumor producing insulin-like growth factor I in a patient with hypoglycemia. Pathol Res Pract 1991;187:114–6.

104. Suh YL, Choi WJ. Benign cystic mesothelioma of the peritoneuмùa case report. J Korean Med Sci 1989;4:111–5.

105. Tahlan RN, Garg P, Bishnoi PK, Singla SL. Mesenteric lipoma: an unusual cause of small intestinal volvulus. Indian J Gastroenterol 1997;16:159.

106. Takenouchi Y, Oda K, Takahara O, et al. Report of a case of benign cystic mesothelioma. Am J Gastroenterol 1995;90:1165–7.

107. Tobioka H, Manabe K, Matsuoka S, Sano F, Mori M. Multicystic mesothelioma of the spermatic cord. Histopathology 1995;27:479–81.

108. Urschel JD, Brooks JS, Werness BA, Antkowiak JG, Takita H. Metachronous benign solitary fibrous tumours of the pleura (localized "mesotheliomas"): a case report. Can J Surg 1998;41:467–9.

109. van de Rijn M, Lombard CM, Rouse RV. Expression of CD34 by solitary fibrous tumors of the pleura, mediastinum, and lung. Am J Surg Pathol 1994;18:814–20.

110. Van Dorpe J, Ectors N, Geboes K, D'Hoore A, Sciot R. Is calcifying fibrous pseudotumor a late sclerosing stage of inflammatory myofibroblastic tumor? Am J Surg Pathol 1999;23:329–35.

111. Weiss SW, Tavassoli FA. Multicystic mesothelioma. An analysis of pathologic findings and biologic behavior in 37 cases. Am J Surg Pathol 1998;12:737–46.

112. Witkin GB, Rosai J. Solitary fibrous tumor of the mediastinum. A report of 14 cases. Am J Surg Pathol 1989;13:547–57.

113. Yesner R, Hurwitz A. Localized pleural mesothelioma of epithelial type. J Thorac Surg 1953;26:325–9.

114. Young RH, Clement PB, McCaughey WT. Solitary fibrous tumors ('fibrous mesotheliomas') of the peritoneum. A report of three cases and a review of the literature. Arch Pathol Lab Med 1990;114:493–5.

115. Zamecnik M, Gomolcak P. Composite multicystic mesothelioma and adenomatoid tumor of the ovary: additional observations suggesting common histogenesis of both lesions. Cesk Patol 2000;36:160–2.

116. Zanetti G. Benign lipoblastoma: first case report of a mesenteric origin. Tumori 1988;74:495–8.

7 LOCALIZED MALIGNANT TUMORS OF THE SEROSAL MEMBRANES

MALIGNANT VERSION OF SOLITARY FIBROUS TUMOR

As noted in chapter 6, solitary fibrous tumors (SFTs) are uncommon in general but are nonetheless the most frequent primary benign neoplasm of the pleura. Rare malignant variants of SFT occur (10,14,16,20,24,34,42,46,59,65,76,85, 117,126,128). *Malignant SFT* can recur, metastasize, and result in the death of the patient. They are often larger tumors than benign SFT, and grossly, may be invasive (fig. 7-1) or become necrotic. Microscopically, malignant SFTs may exhibit recognizable SFT histologic features but with increased mitoses, cytologic atypia, and necrosis (figs. 7-2–7-5). In other examples, malignant SFTs consist of benign-appearing areas mixed with areas of frank sarcoma that do not resemble SFT histologically. Apart from tumors with frankly malignant behavior and sarcomatous histopathologic features, the clinical and morphologic criteria for predicting whether a SFT will behave in a malignant manner or impart a poor prognosis are not well established in the literature.

In their series of 8 new cases of SFT and review of 360 cases from the literature, Briseli et al. (14) concluded that 12 percent of SFTs caused the death of patients. These authors believed that attachment to a pedicle was the best indicator of a favorable prognosis and that nuclear pleomorphism and high mitotic rate are seen in larger tumors but do not necessarily indicate a poor prognosis for those with circumscribed tumors. They also concluded that no single histologic feature was predictive of prognosis. Gold et al. (46) noted poorer survival in patients with SFTs with positive surgical margins, size greater than 10 cm, or a malignant histopathologic component. Yokoi et al. (128) found that their high-grade SFTs did not express CD34, and that p53 expression was associated with fatal outcome, clinical recurrence, nuclear atypia, high mitotic rate, or local invasion. Brozzetti et al. (16) reported that high cellularity and microvessel density, high expression of Ki-67 and CD31, and negativity for CD34 suggested a poor prognosis. In a series of 60 cases of SFT, Magdeleinat et al. (76) classified SFT as malignant if

Figure 7-1

MALIGNANT SOLITARY FIBROUS TUMOR

This cross section shows a lobulated, firm, tan-white tumor with attached pleura (lower right margin).

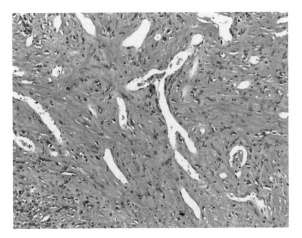

Figure 7-2

MALIGNANT SOLITARY FIBROUS TUMOR

Branching vessels and a pattern that might be seen in an ordinary benign solitary fibrous tumor (SFT).

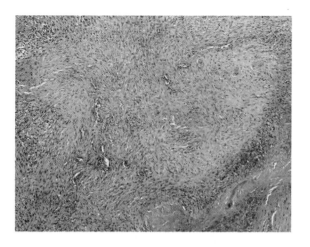

Figure 7-3

MALIGNANT SOLITARY FIBROUS TUMOR

More cellular but cytologically innocuous area from the tumor shown in figure 7-2.

Figure 7-4

MALIGNANT SOLITARY FIBROUS TUMOR

Low-power microscopic view of another area of the tumor shown in figure 7-2 demonstrates a typical SFT pattern on the left, a more cellular pattern with atypical cells in the middle, and an area of necrosis on the right.

Figure 7-5

MALIGNANT SOLITARY FIBROUS TUMOR

Higher-power views of the tumor shown in figure 7-2 demonstrate variable degrees of cytologic atypia. A sarcomatous focus is seen in C.

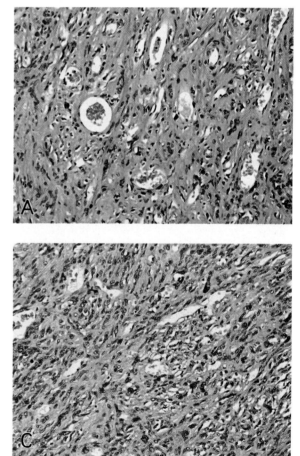

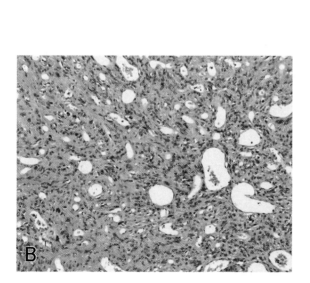

there was at least one of the following histologic criteria: 1) high mitotic activity; 2) high cellularity with crowding and overlapping of nuclei; 3) presence of necrosis; or 4) pleomorphism. Based on these criteria, 22 of the 60 cases were considered malignant. Actuarial 5- and 10-year survival rates for patients with SFTs classified as benign were 97 percent versus 89 percent for those with tumors classified as malignant.

The clinical and histologic features statistically associated with malignancy in SFT have been summarized by Churg (24) as follows: 1) symptoms at presentation (shortness of breath, chest pain, pleural effusion); 2) invasion of adjacent structures; 3) recurrence after resection; 4) attachment to parietal pleura, fissure, or mediastinum, or location in the lung; 5) sessile tumor and size greater than 10 cm; 6) gross findings of hemorrhage and necrosis; 7) microscopic features of cellularity, cytological atypia, mitoses more than 4 mitoses per 10 high-power fields; and 8) overt microscopic foci of sarcoma.

LOCALIZED MALIGNANT MESOTHELIOMA OF PLEURA OR PERITONEUM

Localized malignant mesothelioma is a rare neoplasm that occurs as a discrete, solitary, circumscribed mass usually arising from the visceral or parietal pleura (24,28,39,47,53,78,87,88, 118). Localized malignant mesothelioma has been reported in men and women about equally and patients have an age range from the 40s to the 70s. The tumors are sessile or pedunculated and range in size from a few to 10 cm.

Grossly, localized malignant mesotheliomas appear as circumscribed nodules (fig. 7-6). The histopathology, immunostaining characteristics, and ultrastructure are the same as those of diffuse malignant mesothelioma, including epithelial, sarcomatous, and biphasic types (figs. 7-7–7-11). Localized malignant mesotheliomas are immunopositive for keratins and other markers typically associated with mesotheliomas, and are immunonegative for carcinoma markers (fig. 7-12) (see chapter 4).

Unlike diffuse malignant mesotheliomas, localized malignant mesotheliomas do not spread over the pleura and in some cases, can be successfully excised (with some patients apparently having no recurrence). In other cases, however, localized malignant mesotheliomas recur after

Figure 7-6

LOCALIZED MALIGNANT MESOTHELIOMA

Cut surface of a well-circumscribed mass on the visceral pleura.

surgery and may metastasize. In the original report of localized malignant mesothelioma, Crotty et al. (28) reported six cases that were treated with surgical excision. Of these, three patients had disease-free survival for an extended time after excision whereas the other three had local recurrence of their disease and died within 2 years of initial excision.

Some diffuse malignant mesotheliomas present as a dominant mass, along with diffuse disease that may not always be readily apparent. Thus, accurate diagnosis of localized malignant mesothelioma requires information beyond that obtainable from microscopic slides; in particular, there needs to be evidence that the process is purely localized. This information can be obtained from radiographic examination, which typically shows a sharply circumscribed,

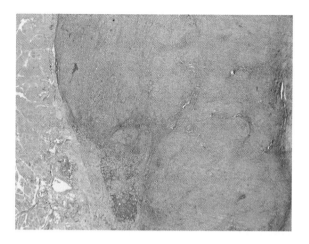

Figure 7-7

LOCALIZED MALIGNANT MESOTHELIOMA

Low-power view shows that the margin is sharply demarcated from diaphragmatic muscle on the left.

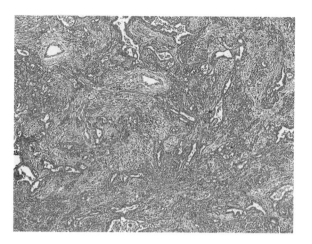

Figure 7-8

LOCALIZED MALIGNANT MESOTHELIOMA

Tubular structures have a histologic pattern similar to that of diffuse malignant mesothelioma.

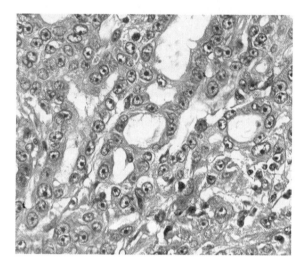

Figure 7-9

LOCALIZED MALIGNANT MESOTHELIOMA

Tubules are lined by mesothelial-like malignant cells with vesicular nuclei, prominent nucleoli, and relatively abundant cytoplasm, identical to the histopathology of some epithelial diffuse malignant mesotheliomas.

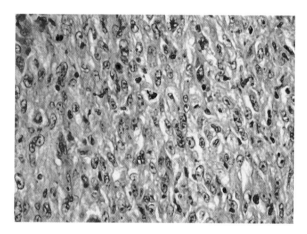

Figure 7-10

LOCALIZED MALIGNANT MESOTHELIOMA

The sarcomatous histologic pattern is identical to that seen in some sarcomatous diffuse malignant mesotheliomas.

DESMOPLASTIC SMALL ROUND CELL TUMOR

Desmoplastic small round cell tumor (DSRCT) is a rare primitive malignancy included in the category of pediatric small round blue cell tumors along with Wilms' tumor and the Ewing's tumor family (Ewing's sarcoma, peripheral primitive neuroectodermal tumor [PNET], Askin's tumor, and others). Classically, DSRCT occurs as an abdominal, pelvic, or paratesticular cancer in adolescent and young adult males aged 15 to 35

pleural-based nodule and no evidence of diffuse pleural thickening or diffuse pleural nodularity; from the surgeon's report of the findings of a solitary nodule and no evidence of diffuse tumor at thoracotomy; or from the gross specimen.

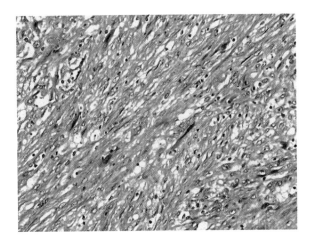

Figure 7-11

LOCALIZED MALIGNANT MESOTHELIOMA

High-power microscopic view shows a sarcomatous pattern identical to that seen in some sarcomatous epithelial diffuse malignant mesotheliomas.

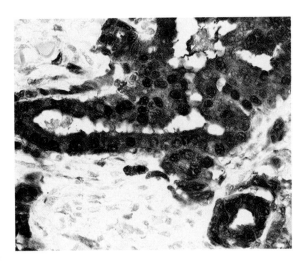

Figure 7-12

LOCALIZED MALIGNANT MESOTHELIOMA

Calretinin immunostain shows nuclear positivity similar to the expected reaction in an epithelial diffuse malignant mesothelioma.

years, typically spreading on the peritoneal surfaces in a fashion similar to diffuse malignant mesothelioma, and presenting with ascites (2,4, 8,9,11,12,15,17,19,21,27,29,32,33,37,38,40,41, 45,51,55,61,63,68,70,72,74,75,82,83,90–92,94,96, 98,102–106,110,111,115,123). In a series from Memorial Sloan-Kettering Cancer Center (46), 90 of 109 cases of DSRCT (83 percent) occurred in males and 103 (94 percent) were intra-abdominal, with an age range of 6 to 49 years and a mean of 22 years. There have been five cases of DSRCT reported in the pleura (19,94). All but one of these were in young adult males (the exception being a teenaged girl); 4 of 5 presented with pleural effusions and 3 of 5 presented with nodular masses encasing the lung, one as studding of the pleural surface and one as a pedunculated mass. DSRCT may also arise in the peritoneum and invade the pleura. Tumors in older patients, females of various ages, and primary in nonserosal tissues have been reported (9,83,90,104,110,123).

Although there are reports of response to aggressive multimodality therapy, DSRCT is generally a very aggressive tumor causing a fatal outcome (33,68,70). In a review of 40 patients with DSRCT from Memorial Sloan-Kettering Cancer Center (103), 29 percent were alive at 3 years from diagnosis. In a series of 35 patients with DSRCT from M.D. Anderson Cancer Center (92),

71 percent were dead of widespread metastases at a mean of 25.2 months from the diagnosis; the remainder were alive with disease.

DSRCTs have a distinctive histopathology consisting of cords and nests of primitive-appearing, small, round, malignant cells within a fibrous stroma (figs. 7-13–7-16). Often intermingled with the small round cells are larger cells with rhabdoid features, including more abundant cytoplasm and eccentric nuclei. Multinucleated sarcomatoid cells, large epithelioid cells with foci of anaplasia, signet ring cells, and other histopathologic variants have been reported (92,96). Immunohistochemistry shows epithelial, mesenchymal, and neural differentiation. Tumors are generally, but not always, immunopositive for desmin, vimentin, keratin, epithelial membrane antigen, Wilms' tumor antigen 1 (WT1), and neuron-specific enolase (11,21,51,91,92,96).

Analogous to cancers in the Ewing's tumor family, DSRCTs have a characteristic chromosomal translocation: t(11;22)(p13;q12) (8,15, 17,38,45,61,63,72,98,103). This translocation produces a chimeric or fusion transcript between the Ewing's sarcoma (*EWS*) gene and the Wilms' tumor (*WT1*) gene and can be detected by reverse transcriptase-polymerase chain reaction (RT-PCR) for diagnostic purposes. Although

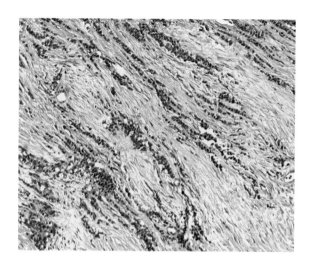
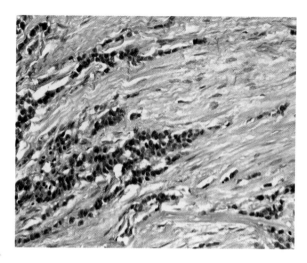

Figure 7-13

DESMOPLASTIC SMALL ROUND CELL TUMOR

Left: Small round blue cells with scant cytoplasm are seen in a background of desmoplasia.
Right: High-power view.

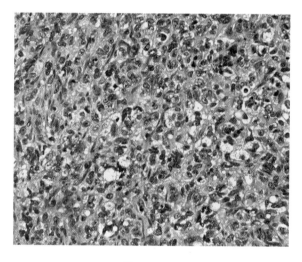

Figure 7-14

DESMOPLASTIC SMALL ROUND CELL TUMOR

Sheets of small round to polygonal cells with scant cytoplasm are present in desmoplastic small round cell tumor (DSRCT).

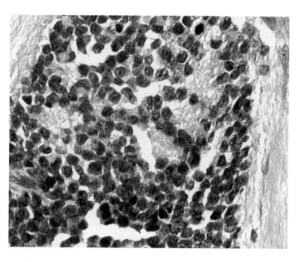

Figure 7-15

DESMOPLASTIC SMALL ROUND CELL TUMOR

High-power microscopy shows small round to polygonal cells that mimic those of small cell carcinoma.

some investigators have proposed that DSRCTs are primitive mesothelial tumors of the serosal membranes or "mesoblastomas," this has not been proven and others challenge this concept.

ASKIN'S TUMOR

The Ewing's sarcoma family of pediatric tumors is defined by the presence of chromosomal translocations that result in gene fusions between the *EWS* gene and a member of the ETS family of transcription factors. *Askin's tumor* is a member of the Ewing's sarcoma family, which also includes peripheral primitive neuroectodermal tumor and neuroepithelioma. Askin's tumors, like the other rare members of the Ewing's sarcoma family of tumors, are small round blue cell tumors of children, adolescents, and young adults. They and other members of the Ewing's

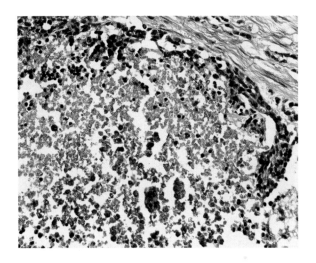

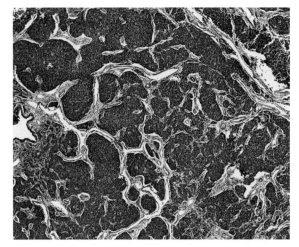

Figure 7-16

DESMOPLASTIC SMALL ROUND CELL TUMOR

Tumor necrosis is present.

sarcoma family of tumors share many common features. Askin's tumors are specifically recognized as aggressive thoracopulmonary tumors that may enter into the differential diagnosis of pleural tumors (1,18,23,31,67,107,108,112, 122). Prognosis is poor.

Askin's tumor presents as a single mass or multiple nodules involving the chest wall and adjacent tissues like the pleura. Histologically, the tumor consists of nests of undifferentiated small round blue cells that may be arranged in rosettes (figs. 7-17, 7-18). Positive immuno-staining for synaptophysin and S-100 protein confirms its neuroectodermal phenotype, but chromogranin immunostaining is often negative and neurosecretory granules are sparse on electron microscopy.

PLEUROPULMONARY BLASTOMA

Pleuropulmonary blastoma (PPB) of childhood is an entity distinct from the biphasic pulmonary blastomas in the lungs of adults and the well-differentiated fetal adenocarcinomas, which were once included in the category of pulmonary blastomas (25,35,77). PPBs are rare embryonal mesenchymal cancers that arise in the lung, pleura, or mediastinum of children, and lack the carcinomatous component seen in adult-type pulmonary blastomas (25,35,36,44, 48–50,52,54,60,69,73,77,80,86,93,95,99–101, 109,119,125,127,129). Thus, PPBs are dysembry-

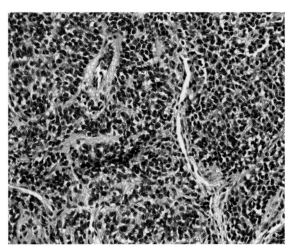

Figure 7-17

ASKIN'S TUMOR

Top: Nests of small blue cells are seen.
Bottom: Higher-power view.

onic or dysontogenetic tumors that are similar to Wilms' tumor, embryonal rhabdomyo-sarcoma, neuroblastoma, and hepatoblastoma. PPBs are often associated with cystic dysplastic diseases of the lung, including cystic adenoma-toid malformation and congenital lung cysts. An association with familial disease and a few characteristic genetic abnormalities, particularly polysomy 18, have been proposed (69,93, 101,109,119,127).

These tumors are usually unilateral (99). Most patients are young children and tumors develop in both sexes (52). Respiratory distress is the

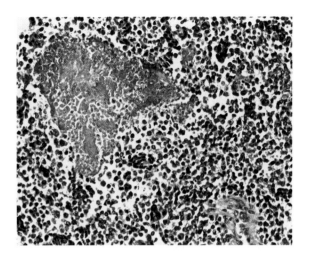

Figure 7-18

ASKIN'S TUMOR

Tumor necrosis is seen.

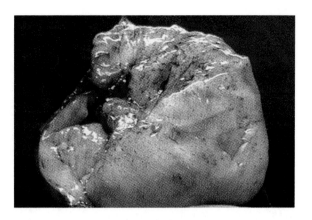

Figure 7-19

PLEUROPULMONARY BLASTOMA

The partially cystic tumor has prominent solid sarcoma-like regions and multiple nodular areas in the cystic component. (Courtesy of Drs. M. Dishop and C. Langston, Houston, TX.)

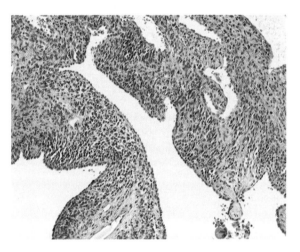

Figure 7-20

PLEUROPULMONARY BLASTOMA

Primitive mesenchymal tissue and cystic spaces.

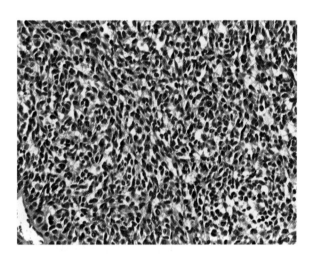

Figure 7-21

PLEUROPULMONARY BLASTOMA

Primitive mesenchymal tissue consists of sheets of small elongated cells with dark nuclei.

most common clinical presentation. PPBs may also mimic empyemas clinically and radiographically, and patients have been reported to present with spontaneous pneumothorax (49,60,71,80). Although outcome is often poor, long-term survival after surgery or multimodal therapy has been reported (54,95,129).

PPBs may be purely cystic (type I), cystic and solid (type II), or purely solid (type III). Clini-

cally, there appears to be a better outcome with the purely cystic type I tumors. Grossly, solid and/or cystic regions of the neoplasm may be observed (fig. 7-19). Histologically, PPBs are generally composed of two cell types: 1) small, primitive, round blastemal cells with hyperchromatic nuclei and scanty cytoplasm and (2) spindle-shaped mesenchymal cells (figs. 7-20, 7-21). Primitive cartilage may be a component

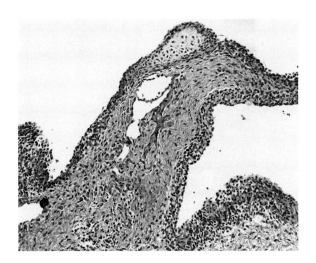 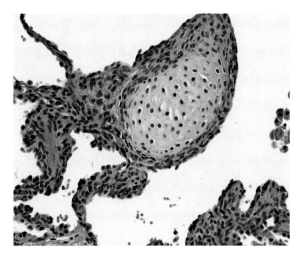

Figure 7-22

PLEUROPULMONARY BLASTOMA

Left: A focus of primitive cartilage and a double cambium layer are in the wall of a cystic space.
Right: Higher-power view of a focus of primitive cartilage.

(fig. 7-22). Sarcomatous areas may be observed, including chondrosarcoma, liposarcoma, and particularly rhabdomyosarcoma; some tumors are predominantly composed of embryonal rhabdomyosarcoma. A neoplastic epithelial component is not present, although entrapped benign mesothelial cells or benign pulmonary epithelial cells may be observed. PPBs are immunopositive for vimentin and other mesenchymal markers according to their differentiation (desmin, HHF35, S-100 protein) and are immunonegative for cytokeratin and epithelial membrane antigen, although entrapped benign mesothelial or epithelial cells may stain for the latter two markers.

LIPOSARCOMA

Primary liposarcomas of the pleura are rare (30, 79,89,116,124). Most occur in older men, but they also occur in women, and the age range in the literature is wide. *Myxoid liposarcoma* is said to be the most common subtype, but all histologic variants of liposarcoma occur (figs. 7-23–7-26). Surgical resection, with or without chemotherapy, is the most common treatment and has mixed success. Liposarcomatous differentiation has occasionally been reported in diffuse malignant mesothelioma and in malignant solitary fibrous tumor (10,66).

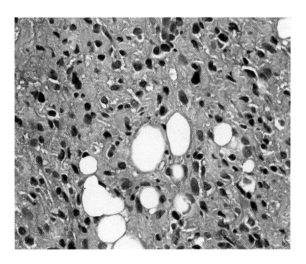

Figure 7-23

LIPOSARCOMA

Dedifferentiated liposarcoma shows poorly differentiated sarcoma cells and cells with cytoplasmic fat vacuoles.

LYMPHOMA

Primary lymphomas of the serosal membranes and cavities are herpesvirus-associated pleomorphic B-cell lymphomas (3,5–7,13,22,26, 43,56–58,64,79,81,97,114,120,121). *Body cavity-based lymphomas* (BCBLs) and *primary effusion lymphomas* (PELs) occur primarily in human

127

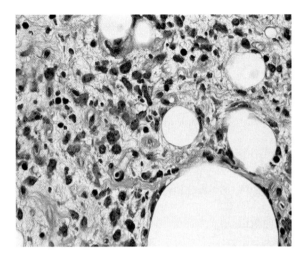

Figure 7-24

LIPOSARCOMA

Liposarcoma shows sarcomatous cells and cells with cytoplasmic fat vacuoles within a myxoid stroma.

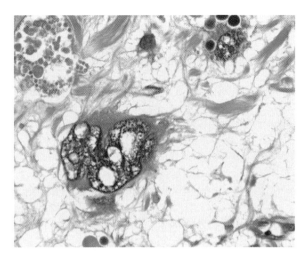

Figure 7-26

LIPOSARCOMA

A highly atypical cell from a pleomorphic liposarcoma displays a multilobulated nucleus.

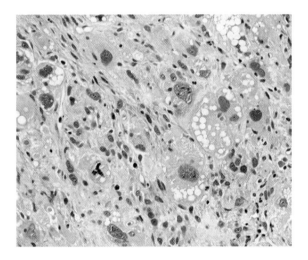

Figure 7-25

LIPOSARCOMA

Pleomorphic liposarcoma shows anaplastic cells, including cells with cytoplasmic fat vacuoles. A cell with an atypical mitosis also has cytoplasmic fat vacuoles.

immunodeficiency virus (HIV)-positive patients. Body cavities are involved without mass lesions and with little or no dissemination outside of the body cavities. BCBLs are consistently associated with DNA from the Kaposi's sarcoma–associated herpesvirus (KSHV), also known as human herpesvirus 8 (HHV-8), a member of the lymphotropic herpes family (13,43,58,114,

121). Epstein-Barr virus (EBV) DNA, another member of the herpes family, is also frequently found in the same lymphoma and it has been postulated by some investigators that HHV-8 and EBV might co-operate in the pathogenesis of BCBL (114). Prognosis is poor.

Morphologically, BCBLs are high-grade, pleomorphic, immunoblastic neoplasms consisting of large round lymphoid cells with large round nuclei, prominent nucleoli, and abundant basophilic cytoplasm of late B-cell phenotype and genotype (figs. 7-27, 7-28). Mitoses are numerous. Some cells may have a plasmacytoid appearance.

Pyothorax-associated lymphomas (PALs) occur in the pleura of patients who have had pyothorax for several decades, a result of artificial pneumothorax for the treatment of pulmonary tuberculosis or tuberculous pleuritis. The majority of cases of PAL have been reported from Japan where the entity was first described. PALs are consistently associated with EBV DNA but only rarely has HHV-8 DNA been reported and, therefore, PAL appears to have a different pathogenesis from BCBL (26,57,114). PALs consist of large lymphoid cells or immunoblasts that have a high mitotic rate. Virtually all are B-cell lymphomas. EBNA-1 and LMP-1 expression can be detected by immunostaining and EBER-1 and EBER-2 RNA by in situ hybridization in the lymphoma cells.

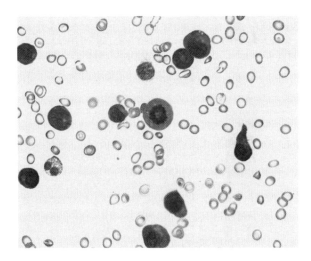

Figure 7-27

PLEURAL EFFUSION LYMPHOMA

Atypical lymphoid cells with a mitosis in the pleural fluid.

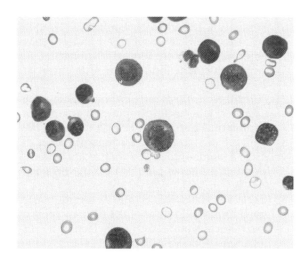

Figure 7-28

PLEURAL EFFUSION LYMPHOMA

Diff-Quik stain demonstrates atypical lymphoid cells.

REFERENCES

1. Aggarwal M, Lakhhar B, Aggarwal BK, Anugu R. Askin tumor: a malignant small cell tumor. Indian J Pediatr 2000;67:853–5.

2. Ali SZ, Nicol TL, Port J, Ford G. Intraabdominal desmoplastic small round cell tumor: cytopathologic findings in two cases. Diagn Cytopathol 1998;18:449–52.

3. Aozasa K. Pyothorax-associated lymphoma. Int J Hematol 1996;65:9–16.

4. Ariffin H, Ariffin WA, Wong KT, Ramanujam TM, Lin HP. Intraabdominal desmoplastic small round cell tumour in an 11-year-old boy. Singapore Med J 1997;38:169–71.

5. Aruga T, Itami J, Nakajima K, et al. Treatment for pyothorax-associated lymphoma. Radiother Oncol 2000;56:59–63.

6. Ascani S, Piccioli M, Poggi S, et al. Pyothorax-associated lymphoma: description of the first two cases detected in Italy. Ann Oncol 1997;8: 1133–8.

7. Ascoli V, Lo-Coco F. Body cavity lymphoma. Curr Opin Pulm Med 2002;8:317–22.

8. Athale UH, Shurtleff SA, Jenkins JJ, et al. Use of reverse transcriptase polymerase chain reaction for diagnosis and staging of alveolar rhabdomyosarcoma, Ewing sarcoma family of tumors, and desmoplastic small round cell tumor. J Pediatr Hematol Oncol 2001;23:99–104.

9. Backer A, Mount SL, Zarka MA, et al. Desmoplastic small round cell tumour of unknown primary origin with lymph node and lung metastases: histological, cytological, ultrastructural, cytogenetic and molecular findings. Virchows Arch 1998;432:135–41.

10. Bai H, Aswad BI, Gaissert H, Gnepp DR. Malignant solitary fibrous tumor of the pleura with liposarcomatous differentiation. Arch Pathol Lab Med 2001;125:406–9.

11. Barnoud R, Sabourin JC, Pasquier D, et al. Immunohistochemical expression of WT1 by desmoplastic small round cell tumor: a comparative study with other small round cell tumors. Am J Surg Pathol 2000;24:830–6.

12. Bisogno G, Roganovich J, Sotti G, et al. Desmoplastic small round cell tumour in children and adolescents. Med Pediatr Oncol 2000;34:338–42.

13. Boshoff C, Chang Y. Kaposi's sarcoma-associated herpesvirus: a new DNA tumor virus. Annu Rev Med 2001;52:453–70.

14. Briselli M, Mark EJ, Dickersin GR. Solitary fibrous tumors of the pleura: eight new cases and review of 360 cases in the literature. Cancer 1981;47:2678–89.

15. Brodie SG, Stocker SJ, Wardlaw JC, et al. EWS and WT-1 gene fusion in desmoplastic small round cell tumor of the abdomen. Hum Pathol 1995;26:1370–4.

16. Brozzetti S, D'Andrea N, Limiti MR, Pisanelli MC, De Angelis R, Cavallaro A. Clinical behavior of solitary fibrous tumors of the pleura. An immunohistochemical study. Anticancer Res 2000;20:4701–6.

17. Burchill SA. Ewing's sarcoma: diagnostic, prognostic, and therapeutic implications of molecular abnormalities. J Clin Pathol 2003;56:96–102.

18. Cabezali R, Lozano R, Bustamante E, et al. Askin's tumor of the chest wall: a case report in an adult. J Thorac Cardiovasc Surg 1994;107:960–2.

19. Cagle PT, Hicks MJ, Haque A. Desmoplastic small round cell tumor of the pleura: report of two new cases and review of the literature. Histopathology 2002;41(Suppl 2):164–8.

20. Chang ED, Lee EH, Won YS, Kim JM, Suh KS, Kim BK. Malignant solitary fibrous tumor of the pleura causing recurrent hypoglycemia; immunohistochemical stain of insulin-like growth factor 1 receptor in three cases. J Korean Med Sci 2001;16:220–4.

21. Charles AK, Moore IE, Berry PJ. Immunohistochemical detection of the Wilms' tumour gene WT1 in desmoplastic small round cell tumour. Histopathology 1997;30:312–4.

22. Cheung C, Schonell M, Manoharan A. A variant of pyothorax-associated lymphoma. Postgrad Med J 1999;75:613–4.

23. Christiansen S, Semik M, Dockhorn-Dworniczak B, et al. Diagnosis, treatment and outcome of patients with Askin-tumors. Thorac Cardiovasc Surg 2000;48:311-5.

24. Churg AM. Localized pleural tumors. In: Cagle PT, ed. Diagnostic pulmonary pathology. New York: M Dekker; 2000:719–35.

25. Cohen M, Emms M, Kaschula RO. Childhood pulmonary blastoma: a pleuropulmonary variant of the adult-type pulmonary blastoma. Pediatr Pathol 1991;11:737–49.

26. Copie-Bergman C, Niedobitek G, Mangham DC, et al. Epstein-Barr virus in B-cell lymphomas associated with chronic suppurative inflammation. J Pathol 1997;183:287–92.

27. Crapanzano JP, Cardillo M, Lin O, Zakowski MF. Cytology of desmoplastic small round cell tumor. Cancer 2002;96:21–31.

28. Crotty TB, Myers JL, Katzenstein AL, Tazelaar HD, Swensen SJ, Churg A. Localized malignant mesothelioma. A clinicopathologic and flow cytometric study. Am J Surg Pathol 1994;18:357–63.

29. Cummings OW, Ulbright TM, Young RH, Del Tos AP, Fletcher CD, Hull MT. Desmoplastic small round cell tumors of the paratesticular region. A report of six cases. Am J Surg Pathol 1997;21:219–25.

30. D'Ambrosio V. First case of liposarcoma from the parietal pleura. J Med Soc N J 1974;71:17–9.

31. Dang NC, Siegel SE, Phillips JD. Malignant chest wall tumors in children and young adults. J Pediatr Surg 1999;34:1773–8.

32. De Alava E, Pardo J. Ewing tumor: tumor biology and clinical applications. Int J Surg Pathol 2001;9:7–17.

33. De Lena M, Caruso ML, Marzullo F, et al. Complete response to chemotherapy in intra-abdominal desmoplastic small round cell carcinoma. A case report. Tumori 1998;84:412–6.

34. de Perrot M, Kurt AM, Robert JH, Borisch B, Spiliopoulos A. Clinical behavior of solitary fibrous tumors of the pleura. Ann Thorac Surg 1999;67:1456–9.

35. Dehner LP. Pleuropulmonary blastoma is THE pulmonary blastoma of childhood. Semin Diagn Pathol 1994;11:144–51.

36. Delahunt B, Thomson KJ, Ferguson AF, Neale TJ, Meffan PJ, Nacey JN. Familial cystic nephroma and pleuropulmonary blastoma. Cancer 1993;71:1338–42.

37. Devaney K. Intra-abdominal desmoplastic small round cell tumor of the peritoneum in a young man. Ultrastruct Pathol 1994;18:389–98.

38. Dockhorn-Dworniczak B, Schafer KL, Blasius S, et al. Assessment of molecular genetic detection of chromosome translocations in the differential diagnosis of pediatric sarcomas. Klin Padiatr 1997;209:156–64.

39. Erkilic S, Sari I, Tuncozgur B. Localized pleural malignant mesothelioma. Pathol Int 2001;51:812–5.

40. Ferlicot S, Coue O, Gilbert E, et al. Intraabdominal desmoplastic small round cell tumor: report of a case with fine needle aspiration, cytologic diagnosis and molecular confirmation. Acta Cytol 2001;45:617–21.

41. Frappaz D, Bouffet E, Dolbeau D, et al. Desmoplastic small round cell tumors of the abdomen. Cancer 1994;73:1753–6.

42. Fukasawa Y, Takada A, Tateno M, et al. Solitary fibrous tumor of the pleura causing recurrent hypoglycemia by secretion of insulin-like growth factor II. Pathol Int 1998;48:47–52.

43. Gaidano G, Castanos-Velez E, Biberfeld P. Lymphoid disorders associated with HHV-8/KSHV infection: facts and contentions. Med Oncol 1999;16:8–12.

44. Gelven PL, Hopkins MA, Green CA, Harley RA, Wilson MM. Fine-needle aspiration cytology of pleuropulmonary blastoma: case report and review of the literature. Diagn Cytopathol 1997;16:336–40.

45. Gerald WL, Ladanyi M, de Alava E, et al. Clinical, pathologic and molecular spectrum of tumors associated with t(11;22)(p13;q12): desmoplastic small round cell tumor and its variants. J Clin Oncol 1998;16:3028–36.
46. Gold JS, Antonescu CR, Hajdu C, et al. Clinicopathologic correlates of solitary fibrous tumors. Cancer 2002;94:1057–68.
47. Gomez-Roman JJ, Mons-Lera R, Olmedo IS, Val-Bernal JF. Flow cytometric analysis of a localized malignant mesothelioma. Ann Thorac Surg 2002;73:1292–4.
48. Granata C, Gambini C, Carlini C, et al. Pleuropulmonary blastoma. Eur J Pediatr Surg 2001; 11:271–3.
49. Guler E, Kutluk MT, Yalcin B, Cila A, Kale G, Buyukpamukcu M. Pleuropulmonary blastoma in a child presenting with pneumothorax. Tumori 2001;87:340–2.
50. Hachitanda Y, Aoyama C, Sato JK, Shimada H. Pleuropulmonary blastoma in childhood. A tumor of divergent differentiation. Am J Surg Pathol 1993;17:382–91.
51. Hill DA, Pfeifer JD, Marley EF, et al. WT1 staining reliably differentiates desmoplastic small round cell tumor from Ewing sarcoma/primitive neuroectodermal tumor. An immunohistochemical and molecular diagnostic study. Am J Clin Pathol 2000;114:345–53.
52. Hill DA, Sadeghi S, Schultz MZ, Burr JS, Dehner LP. Pleuropulmonary blastoma in an adult: an initial case report. Cancer 1999;85:2368–74.
53. Imura J, Ichikawa K, Takeda J, et al. Localized malignant mesothelioma of the epithelial type occurring as a primary hepatic neoplasm: a case report with review of the literature. APMIS 2002;110:789–94.
54. Indolfi P, Casale F, Carli M, et al. Pleuropulmonary blastoma: management and prognosis of 11 cases. Cancer 2000;89:1396–401.
55. Insabato L, DiVizio D, Lambertini M, Bucci L, Pettinato G. Fine needle aspiration cytology of desmoplastic small round cell tumor. A case report. Acta Cytol 1999;43:641–6.
56. Kanno H, Aozasa K. Mechanism for the development of pyothorax-associated lymphoma. Pathol Int 1998;48:653–64.
57. Kanno H, Ohsawa M, Hashimoto M, Iuchi K, Nakajima Y, Aozasa K. HLA-A alleles of patients with pyothorax-associated lymphoma: anti-Epstein-Barr virus (EBV) host immune responses during the development of EBV latent antigen-positive lymphomas. Int J Cancer 1999;82:630–4.
58. Karcher DS, Alkan S. Human herpesvirus-8-associated body cavity-based lymphoma in human immunodeficiency virus-infected patients: a unique B-cell neoplasm. Hum Pathol 1997;28: 801–8.
59. Kashiwabara K, Kishi K, Nakamura H, et al. Malignant solitary fibrous tumor arising in the right buttock associated with metastatic parietal pleural and intrapulmonary tumors in addition to pleural effusion. Intern Med 1997;36:732–7.
60. Katz DS, Scalzetti EM, Groskin SA, Kohman LJ, Patel LS, Landas S. Pleuropulmonary blastoma simulating an empyema in a young child. J Thorac Imaging 1995;10:112–6.
61. Katz RL, Quezado M, Senderowicz AM, Villalba L, Laskin WB, Tsokos M. An intra-abdominal small round cell neoplasm with features of primitive neuroectodermal and desmoplastic round cell tumor and a EWS/FLI-1 fusion transcript. Hum Pathol 1997;28:502–9.
62. Kawa K. Epstein-Barr virus–associated diseases in humans. Int J Hematol 2000;71:108–17.
63. Kim J, Lee K, Pelletier J. The desmoplastic small round cell tumor t(11;22) translocation produces EWS/WT1 isoforms with differing oncogenic properties. Oncogene 1998;16:1973–9.
64. Kinoshita T, Ishii K, Taira Y, Naganuma H. Malignant lymphoma arising from chronic tuberculous empyema. A case report. Acta Radiol 1997;38:833–5.
65. Kishi K, Homma S, Tanimura S, Matsushita H, Nakata K. Hypoglycemia induced by secretion of high molecular weight insulin-like growth factor-II from a malignant solitary fibrous tumor of the pleura. Intern Med 2001;40:341–4.
66. Krishna J, Haqqani MT. Liposarcomatous differentiation in diffuse pleural mesothelioma. Thorax 1993;48:409–10.
67. Kurashima K, Muramoto S, Ohta Y, Fujimura M, Matsuda T. Peripheral neuroectodermal tumor presenting pleural effusion. Intern Med 1994;33:783–5.
68. Kurre P, Felgenhauer JL, Miser JS, Patterson K, Hawkins DS. Successful dose-intensive treatment of desmoplastic small round cell tumor in three children. J Pediatr Hematol Oncol 2000; 22:446–50.
69. Kusafuka T, Kuroda S, Inoue M, et al. P53 gene mutations in pleuropulmonary blastomas. Pediatr Hematol Oncol 2002;19:117–28.
70. Kushner BH, LaQuaglia MP, Wollner N, et al. Desmoplastic small round-cell tumor: prolonged progression-free survival with aggressive multimodality therapy. J Clin Oncol 1996;14: 1526–31.
71. Kuzucu A, Soysal O, Yakinci C, Aydin NE. Pleuropulmonary blastoma: report of a case presenting with spontaneous pneumothorax. Eur J Cardiothorac Surg 2001;19:229–30.

72. Ladanyi M, Gerald W. Fusion of the EWS and WT1 genes in the desmoplastic small round cell tumor. Cancer Res 1994;54:2837–40.

73. Lallier M, Bouchard S, Di Lorenzo M, et al. Pleuropulmonary blastoma: a rare pathology with an even rarer presentation. J Pediatr Surg 1999; 34:1057–9.

74. Liang WY, Chen WY, Tsay SH, Chiang H. Desmoplastic small round cell tumor—report of 3 cases and review of the literature. Kaohsiung J Med Sci 2000;16:261–5.

75. Lippe P, Berardi R, Cappalletti C, et al. Desmoplastic small round cell tumour: a description of two cases and review of the literature. Oncology 2003;64:14–7.

76. Magdeleinat P, Alifano M, Petino A, et al. Solitary fibrous tumors of the pleura: clinical characteristics, surgical treatment and outcome. Eur J Cardiothorac Surg 2002;21:1087–93.

77. Manivel JC, Priest JR, Watterson J, et al. Pleuropulmonary blastoma. The so-called pulmonary blastoma of childhood. Cancer 1988;62:1516–26.

78. Matsukuma S, Aida S, Hata Y, Sugiura Y, Tamai S. Localized malignant peritoneal mesothelioma containing rhabdoid cells. Pathol Int 1996;46: 389–91.

79. McGregor DH, Dixon AY, Moral L, Kanabe S. Liposarcoma of pleural cavity with recurrence as malignant fibrous histiocytoma. Ann Clin Lab Sci 1987;17:83–92.

80. Merriman TE, Beasley SW, Chow CW, Smith PJ, Robertson CF. A rare tumor masquerading as an empyema: pleuropulmonary blastoma. Pediatr Pulmonol 1996;22:408–11.

81. Mukerji PK, Suryakant, Babu KS, Mehrotra P, Agarwal PK, Tandon R. Non-Hodgkin's lymphoma masquerading as empyema thoracis. Indian J Chest Dis Allied Sci 1997;39:259–62.

82. Murosaki N, Matsumiya K, Kokado Y, et al. Retrovesical desmoplastic small round cell tumor in a patient with urinary frequency. Int J Urol 2001;8:245–8.

83. Murray JC, Minifee PK, Trautwein LM, Hicks MJ, Langston C, Morad AB. Intraabdominal desmoplastic small round cell tumor presenting as a gastric mural mass with hepatic metastases. J Pediatr Hematol Oncol 1996;18:289–92.

84. Nicol KK, Geisinger KR. The cytomorphology of pleuropulmonary blastoma. Arch Pathol Lab Med 2000;124:416–8.

85. Nonaka M, Kadokura M, Takaba T. Benign solitary fibrous tumor of the parietal pleura which invaded the intercostals muscle. Lung Cancer 2001;31:325–9.

86. Novak R, Dasu S, Agamanolis D, Herold W, Malone J, Waterson J. Trisomy 8 is a character-istic finding in pleuropulmonary blastoma. Pediatr Pathol Lab Med 1997;17:99–103.

87. Ojeda HF, Mech K Jr, Hicken WJ. Localized malignant mesothelioma: a case report. Am Surg 1998;64:881–5.

88. Okamura H, Kamei T, Mitsuno A, Hongo H, Sakuma N, Ishihara T. Localized malignant mesothelioma of the pleura. Pathol Int 2001;51:654–60.

89. Okby NT, Travis WD. Liposarcoma of the pleural cavity: clinical and pathologic features of 4 cases with a review of the literature. Arch Pathol Lab Med 2000;124:699–703.

90. Ordonez NG. Desmoplastic small round cell tumor: I: a histopathologic study of 39 cases with emphasis on unusual histological patterns. Am J Surg Pathol 1998;22:1303–13.

91. Ordonez NG. Desmoplastic small round cell tumor: II: an ultrastructural and immunohistochemical study with emphasis on new immunohistochemical markers. Am J Surg Pathol 1998;22:1314–27.

92. Ordonez NG, Sahin AA. CA 125 production in desmoplastic small round cell tumor: report of a case with elevated serum levels and prominent signet ring morphology. Hum Pathol 1998;29:294–9.

93. Pacinda SJ, Ledet SC, Gondo MM, et al. p53 and MDM2 immunostaining in pulmonary blastomas and bronchogenic carcinomas. Hum Pathol 1996;27:542–6.

94. Parkash V, Gerald WL, Parma A, Miettinen M, Rosai J. Desmoplastic small round cell tumor of the pleura. Am J Surg Pathol 1995;19:659–65.

95. Parsons SK, Fishman SJ, Hoorntje LE, et al. Aggressive multimodal treatment of pleuropulmonary blastoma. Ann Thorac Surg 2001;72:939–42.

96. Pasquinelli G, Montanaro L, Martinelli GN. Desmoplastic small round-cell tumor: a case report on the large cell variant with immunohistochemical, ultrastructural, and molecular genetic analysis. Ultrastruct Pathol 2000;24:333–7.

97. Perez MT, Cabello-Inchausti B, Viamonte M Jr, Nixon D. Pleural body cavity-based lymphoma. Ann Diagn Pathol 1998;2:127–34.

98. Perez RP, Zhang PJ. Detection of EWS-WT1 fusion mRNA in ascites of a patient with desmoplastic small round cell tumor by RT-PCR. Hum Pathol 1999;30:239–42.

99. Picaud JC, Levrey H, Bouvier R, et al. Bilateral cystic pleuropulmonary blastoma in early infancy. J Pediatr 2000;136:834–6.

100. Priest JR, McDermott MB, Bhatia S, Watterson J, Manivel JC, Dehner LP. Pleuropulmonary blastoma: a clinicopathologic study of 50 cases. Cancer 1997;80:147–61.

101. Priest JR, Watterson J, Strong L, et al. Pleuro-pulmonary blastoma: a marker for familial disease. J Pediatr 1996;128:220–4.
102. Quaglia MP, Brennan MF. The clinical approach to desmoplastic small round cell tumor. Surg Oncol 2000;9:77–81.
103. Rauscher FJ 3rd, Benjamin LE, Fredericks WJ, Morris JF. Novel oncogenic mutations in the WT1 Wilms' tumor suppressor gene: a t(11;22) fuses the Ewing's sarcoma gene, EWS1, to WT1 in desmoplastic small round cell tumor. Cold Spring Harb Symp Quant Biol 1994;59:137–46.
104. Reich O, Justus J, Tamussino KF. Intra-abdominal desmoplastic small round cell tumor in a 68-year-old female. Eur J Gynaecol Oncol 2000;21:126–7.
105. Resnick MB, Donovan M. Intra-abdominal desmoplastic small round cell tumor with extensive extra-abdominal involvement. Pediatr Pathol Lab Med 1995;15:797–803.
106. Roganovich J, Bisogno G, Cecchetto G, D'Amore ES, Carli M. Paratesticular desmoplastic small round cell tumor: case report and review of the literature. J Surg Oncol 1999;71:269–72.
107. Sabate JM, Franquet T, Parellada JA, Monill JM, Oliva E. Malignant neuroectodermal tumour of the chest wall (Askin tumour): CT and MR findings in eight patients. Clin Radiol 1994;49:634–8.
108. Sallustio G, Pirronti T, Lasorella A, Natale L, Bray A, Marano P. Diagnostic imaging of primitive neuroectodermal tumour of the chest wall (Askin tumour). Pediatr Radiol 1998;28:697–702.
109. Sciot R, Dal Cin P, Brock P, et al. Pleuropulmonary blastoma (pulmonary blastoma of childhood): genetic link with other embryonal malignancies? Histopathology 1994;24:559–63.
110. Slomovitz BM, Girotra M, Aledo A, et al. Desmoplastic small round cell tumor with primary ovarian involvement: case report and review. Gynecol Oncol 2000;79:124–8.
111. Syed S, Haque AK, Hawkins HK, Sorensen PH, Cowan DF. Desmoplastic small round cell tumor of the lung. Arch Pathol Lab Med 2002;126:1226–8.
112. Takanami I, Imamura T, Naruke M, Kodaira S. Long-term survival after repeated resections of Askin tumor recurrences. Eur J Cardiothorac Surg 1998;13:313–5.
113. Taneli C, Genc A, Erikci V, Yuce G, Balik E. Askin tumors in children: a report of four cases. Eur J Pediatr Surg 1998;8:312–4.
114. Taniere P, Manai A, Charpentier R, et al. Pyothorax-associated lymphoma: relationship with Epstein-Barr virus, human herpes virus-8 and body cavity-based high grade lymphomas. Eur Respir J 1998;11:779–83.
115. Trupiano JK, Machen SK, Barr FG, Goldblum JR. Cytokeratin-negative desmoplastic small round cell tumor: a report of two cases emphasizing the utility of reverse transcriptase-polymerase chain reaction. Mod Pathol 1999;12:849–53.
116. Urabe M, Mizobuchi N, Funabiki H, Seki E, Okada T, Sakakibara N. A case of liposarcoma originating in the chest wall. Nippon Geka Hokan 1995;64:131–8.
117. Uzoaru I, Chou P, Reyes-Mugica M. Malignant solitary fibrous tumor of the pleura. Pediatr Pathol 1994;14:11–8.
118. Val-Bernal JF, Figols J, Gomez-Roman JJ. Incidental localized (solitary) epithelial mesothelioma of the pericardium: case report and literature review. Cardiovasc Pathol 2002;11:181–5.
119. Vargas SO, Nose V, Fletcher JA, Perez-Atayde AR. Gains of chromosome 8 are confined to mesenchymal components in pleuropulmonary blastoma. Pediatr Dev Pathol 2001;4:434–45.
120. Vince A, Begovac J, Kessler H, et al. AIDS-related body cavity-based lymphoma. A case report. Acta Cytol 2001;45:420–4.
121. Wakely PE Jr, Menezes G, Nuovo GJ. Primary effusion lymphoma: cytopathologic diagnosis using in situ molecular genetic analysis for human herpesvirus 8. Mod Pathol 2002;15:944–50.
122. Watt AJ. Chest wall lesions. Paediatr Respir Rev 2002;3:328–38.
123. Wolf AN, Ladanyi M, Paull G, Blaugrund JE, Westra WH. The expanding clinical spectrum of desmoplastic small round-cell tumor: a report of two cases with molecular confirmation. Hum Pathol 1999;30:430–5.
124. Wong WW, Pluth JR, Grado GL, Schild SE, Sanderson DR. Liposarcoma of the pleura. Mayo Clin Proc 1994;69:882–5.
125. Wright JR Jr. Pleuropulmonary blastoma: A case report documenting transition from type I (cystic) to type III (solid). Cancer 2000;88:2853–8.
126. Yamamoto R, Tada H, Kishi A, Tojo T. Malignant solitary fibrous tumor in the pleura. Jpn J Thorac Cardiovasc Surg 2000;48:736–8.
127. Yang P, Hasegawa T, Hirose T, et al. Pleuropulmonary blastoma: fluorescence in situ hybridization analysis indicating trisomy 2. Am J Surg Pathol 1997;21:854–9.
128. Yokoi T, Tsuzuki T, Yatabe Y, et al. Solitary fibrous tumour: significance of p53 and CD34 immunoreactivity in its malignant transformation. Histopathology 1998;32:423–32.
129. Yusuf U, Dufour D, Jenrette JM 3rd, Abboud MR, Laver J, Barredo JC. Survival with combined modality therapy after intracerebral recurrence of pleuropulmonary blastoma. Med Pediatr Oncol 1998;30:63–6.

8 MISCELLANEOUS CONDITIONS

PARIETAL PLEURAL PLAQUES

Pleural plaques are localized scars that are typically located on the parietal pleura. They are often found on the surface of the diaphragm or posterolaterally, running along the direction of the ribs (42). They are sharply circumscribed, ivory-colored, firm, raised lesions that vary in thickness from a millimeter to more than a centimeter. Their surface may be either smooth or nodular, and in the latter circumstance, has been described as resembling candle wax drippings (fig. 8-1). They are often calcified or even ossified, which renders them readily visible on chest films or computerized tomography (CT). Pleural plaques are associated with asbestos exposure in the vast majority of cases.

Histologically, plaques consist of layers of acellular hyalinized collagen arranged in a "basket-weave" pattern (fig. 8-2) (7,42). Solid variants have also been described, in which the basket-weave pattern is not apparent. A focal collection of lymphocytes is often seen at the interface between the plaque and the adipose tissue of the chest wall. Areas of calcification may be observed. Rarely, a layer of mesothelial cells overlies the serosal surface. Immunohistochemical stains for cytokeratins may highlight occasional spindle cells between the layers of collagen.

Plaques must be distinguished from localized fibrous tumors and the desmoplastic variant of malignant mesothelioma. Plaques are much less cellular than the former, and have a much more regular arrangement than the latter. Whereas the collagen bundles in desmoplastic mesothelioma tend to be arranged in storiform or haphazard patterns, they are generally parallel in plaques. The term "fibrous mesothelioma" should not be applied to pleural plaques, since plaques are benign, reactive, and non-neoplastic. There is no evidence that plaques are a precursor lesion for mesothelioma.

ROUNDED ATELECTASIS

Rounded atelectasis, also known as *folded lung* or *Blesovsky's syndrome*, is a localized area of atelectatic lung parenchyma that has been entrapped by an overlying focus of fibrotic visceral pleura

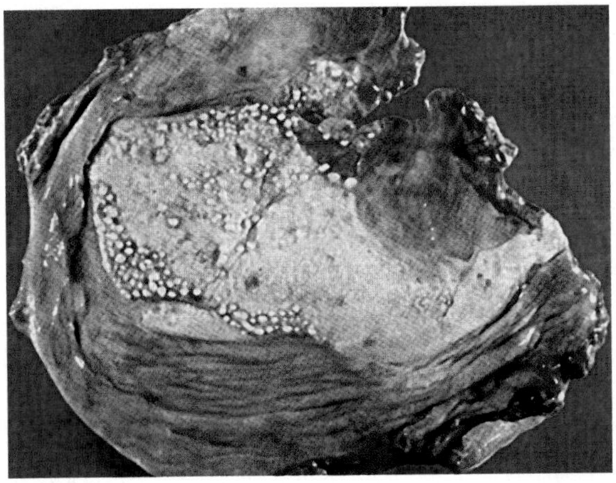

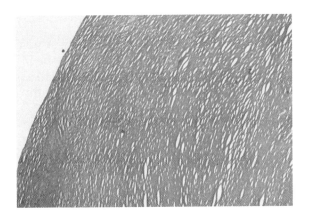

Figure 8-1

PARIETAL PLEURAL PLAQUE

This plaque, located on the surface of the diaphragm, is sharply circumscribed, tan, and focally nodular, giving a "candle-wax dripping" appearance.

Figure 8-2

PARIETAL PLEURAL PLAQUE

The plaque is composed of layers of acellular hyalinized collagen arranged in a "basket-weave" pattern.

135

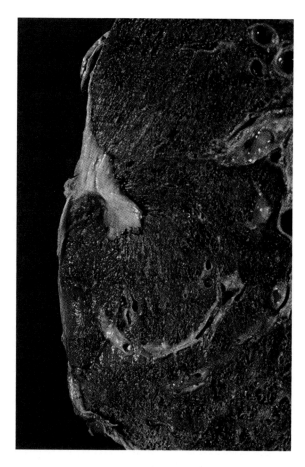

Figure 8-3

ROUNDED ATELECTASIS

There is a localized area of visceral pleural fibrosis that appears to be retracted into the lung parenchyma. The adjacent lung is atelectatic. The patient had prior radiation therapy for Hodgkin's disease.

(fig. 8-3) (42). These lesions typically occur posteriorly in the lower lung zones and may be bilateral. Most cases are caused by exposure to asbestos, but rare cases may be related to hemothorax, pleural infection, radiation, or uremia (27).

Rounded atelectasis has a distinctive radiographic appearance, consisting of a pleural-based mass with overlying pleural thickening. CT scans may show a characteristic "comet-tail" sign, composed of a bronchovascular bundle emanating from the mass and curving towards the hilum of the lung. Other markers of asbestos exposure, such as pleural plaques or calcification, may or may not be present. These lesions occasionally are mistaken clinically for a neoplasm (34).

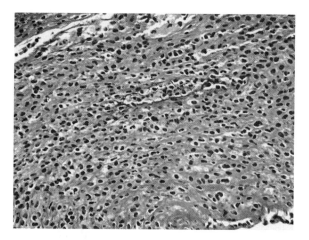

Figure 8-4

HISTIOCYTIC/MESOTHELIAL HYPERPLASIA

Sheets of cells are present, with round to oval nuclei and abundant cytoplasm. The cells resemble mesothelial cells.

Histologically, rounded atelectasis appears as atelectatic lung subjacent to an area of visceral pleural fibrosis. The fibrotic pleura is often puckered inward. Asbestos bodies are sometimes seen in the lung parenchyma, especially when iron stains are performed. A diagnosis of rounded atelectasis should be considered in any case with a pleural-based mass on radiographic studies in which the resected specimen shows only collapsed lung and overlying pleural thickening.

HISTIOCYTIC/MESOTHELIAL HYPERPLASIA

The common reaction to irritation in the serosal membranes is a proliferation of mesothelial cells. A small number of cases, however, have been described as *nodular histiocytic/mesothelial hyperplasia* in transbronchial biopsies (6); *nodular histiocytic hyperplasia* in the pleura, hernia sacs, and lamina propria of the bladder (41); and *mesothelial/monocytic incidental cardiac excrescences* in the heart (52). Ordonez et al. (41) have suggested that the proliferative lesions in hernia sacs described by Rosai et al. (48) also fall into this category.

The common histologic finding in all cases is a proliferation of sheets of fairly monomorphous cells that are much smaller than the usual mesothelial cell. These cells have pale-staining cytoplasm and round to oval nuclei that may be indented (figs. 8-4, 8-5); nucleoli are inconspicuous. Mitoses can be present. In occasional instances, the cytoplasm is vacuolated. The cells are positive for histiocyte markers such as CD68 (fig. 8-6) but are negative for the usual mesothelial markers including keratin and calretinin and for typical carcinoma markers such as carcinoembryonic antigen (CEA). A small numbers of reactive mesothelial cells may be admixed within the lesions.

The major diagnostic problem is that these proliferations may mimic neoplasms, and in fact, occasionally they form grossly visible masses (41). In many instances they appear to be associated with either prior surgery or instrumentation, and sometimes with the presence of a real neoplasm. Histiocytic/mesothelial hyperplasia is a purely benign reactive process.

AMYLOIDOSIS

Pleural involvement is rarely seen in *amyloidosis*, but amyloidosis may be associated with the development of a pleural effusion (29,31). Patients may present with dyspnea and chest pain (1). Diffuse interstitial amyloidosis often involves the underlying lung parenchyma.

The macroscopic appearance is that of inflamed pleura and multiple small nodules (3). Rarely, there may be diffuse pleural thickening that clinically and radiographically resembles mesothelioma (1). The diagnosis may be made by Cope needle biopsy, with appropriate ancillary studies. Amyloid has an amorphous, smudgy appearance on hematoxylin and eosin (H&E)-stained sections, and may be associated with a lymphoplasmacytic infiltrate. Foreign body giant cells may be observed. Amyloid stains with Congo red, and an apple green birefringence upon examination with polarizing microscopy is characteristic. Ultrastructurally,

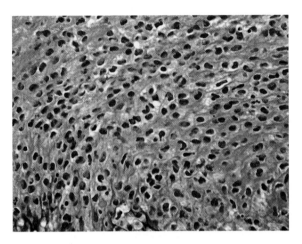

Figure 8-5

HISTIOCYTIC/MESOTHELIAL HYPERPLASIA

Bland nuclear features and poorly defined cytoplasmic borders are seen.

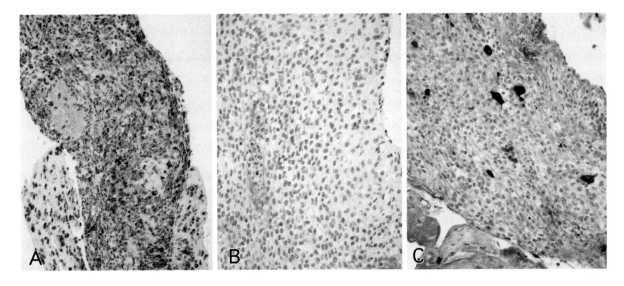

Figure 8-6

HISTIOCYTIC/MESOTHELIAL HYPERPLASIA

The cells stain strongly for CD68, a histiocyte marker (A), but not for the mesothelial markers, calretinin (B) and cytokeratin (C). A few entrapped mesothelial cells stain for cytokeratin (C).

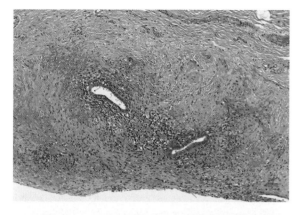

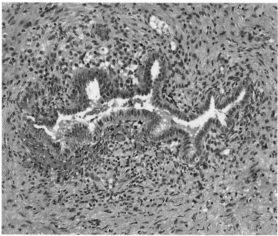

Figure 8-7

ENDOMETRIOSIS

Top: Medium magnification shows an endometrial implant on the pelvic peritoneum that consists of endometrial-type glands and stroma.

Bottom: Higher magnification shows details of endometrial gland morphology.

fibrils with a diameter of 10 nm are seen haphazardly arrayed.

ENDOMETRIOSIS/ENDOSALPINGIOSIS

Endometriosis is a metaplastic condition of the peritoneum that may present as pelvic pain or as an incidental finding at laparotomy. The process may involve the peritoneal and, rarely, the pleural surfaces. The latter has been associated with catamenial pneumothorax. Macroscopically, endometriosis appears as blue or brown nodules or cysts that may be hemorrhagic or fibrotic (17). Extensive adhesions may be present, and in

some cases, fibrosis and distortion are so severe as to cause bowel obstruction.

Histologically, endometriosis is characterized by a benign endometrial-type glandular epithelium in association with endometrial-type stroma (fig. 8-7). Foci of endometriosis may show alterations similar to those of the cycling endometrium, and may even undergo atrophy in postmenopausal women. Hyperplastic and metaplastic changes also occur, and variable degrees of nuclear atypia may be observed in the glandular elements. Necrosis is sometimes observed in association with foamy histiocytes and areas of fibrosis. Smooth muscle is abundant in exceptional cases, and is believed to be derived from müllerian stroma. When prominent, this lesion is referred to as *endomyometriosis*.

Malignant neoplasms have been reported to arise from endometriosis, most commonly, endometrioid or clear cell carcinoma (24). Other reported histologic variants include endometrial stromal sarcoma, malignant mixed mesodermal tumor, pelvic adenosarcoma, and sex cord tumor with annular tubules (10,21,35,50).

Endosalpingiosis, another metaplastic process that can involve the peritoneal cavity, is found mainly in the pelvic peritoneum, omentum, and on ovarian surfaces (4); however, it may involve any area of the peritoneum as well as intra-abdominal lymph nodes. Endosalpingiosis may be an incidental finding at laparotomy or be associated with pelvic pain (25).

Macroscopically, the lesions are typically small and may become cystic. In rare instances, the process may be so extensive as to mimic a neoplastic process (8). Endosalpingiosis has a characteristic microscopic appearance: glands lined by a single layer of bland-appearing ciliated epithelial cells are associated with a fibrotic stroma that lacks features of endometrial-type stroma. In addition, endocervical-type glands may be encountered in the pelvic peritoneum (4). Psammoma bodies may be observed in association with endosalpingiosis (22).

DECIDUOSIS

Deciduosis is an ectopic decidual reaction that occurs in the peritoneal cavity in women during pregnancy or following hormonal therapy. It may be seen in pelvic submesothelial stroma and, less frequently, in abdominal sites remote from the pelvis (54). Deciduosis is usually an incidental microscopic finding, but in some cases may be so florid as to suggest a malignant process (36).

Peritoneal decidual reaction is similar morphologically to that seen in the uterine cervix and oviduct. It consists of sheets of large cells with abundant, pale, granular cytoplasm and bland nuclear features (fig. 8-8). The finding of necrosis and nuclear atypia may cause confusion with deciduoid malignant mesothelioma (39). The nuclear atypia observed in ectopic decidua, however, is usually less severe than that seen in deciduoid mesothelioma. Furthermore, the deciduoid variant of malignant mesothelioma stains strongly for cytokeratins, whereas true decidua is keratin negative.

LEIOMYOMATOSIS PERITONEALIS DISSEMINATA

Another example of metaplasia of submesothelial cells involves the proliferation of smooth muscle cells, referred to as *leiomyomatosis peritonealis disseminata* (33). This condition is associated with typical uterine leiomyomas, and there is also a strong association with pregnancy or oral contraceptive use (13). Rarely, it occurs in association with endometriosis. Because of the disseminated nature of this process, there is the potential for confusion with metastatic leiomyosarcoma.

Macroscopically, the lesion consists of well-demarcated, firm, white nodules. Histologically, it is composed of smooth muscle cells with bland, uniform nuclei (figs. 8-9). The cells stain for desmin, smooth muscle actin, and muscle-specific actin, but are negative for cytokeratins. In addition, there is positive staining for progesterone and estrogen receptors (5). The clinical course is benign, with spontaneous regression in most instances. Recurrence has been reported (13).

CHONDROID AND OSSEOUS METAPLASIA

Osseous or cartilaginous (chondroid) metaplasia rarely involves the peritoneal cavity (14,53). These lesions are usually an incidental finding at laparotomy, and most patients have had a history of prior surgery. Interestingly, cartilaginous metaplasia has been reported in women whereas reported cases of heterotopic mesenteric ossification have occurred primarily in men (14,53).

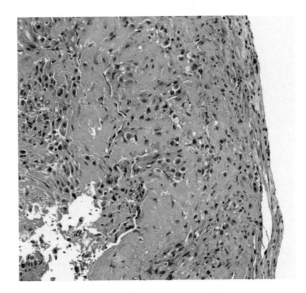

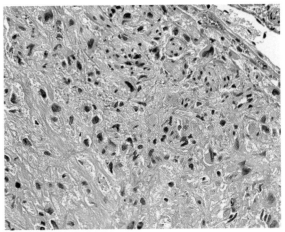

Figure 8-8

DECIDUOSIS

Top: Sheets of large cells in a desmoplastic stroma involve the peritoneal surface.

Bottom: Higher magnification shows decidual cells with abundant eosinophilic cytoplasm, irregular nuclei, and ill-defined cytoplasmic borders. Such an appearance may be confused with malignancy.

Osseous metaplasia consists of single or multiple nodules of well-circumscribed mature bone, while cartilaginous metaplasia consists of nodules of mature cartilage. The lesions range in size from 2 mm to 2 cm in diameter. Histologically, there is no evidence of cytologic atypia, and the lesions are covered by intact mesothelium. These lesions presumably represent metaplasia of the submesothelial mesenchyme.

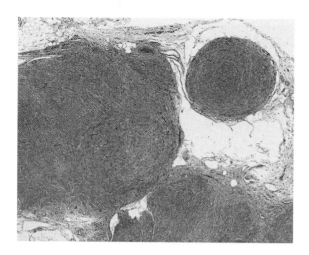 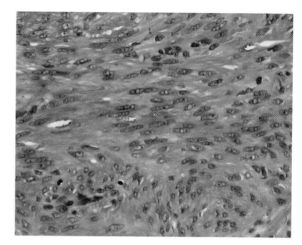

Figure 8-9

LEIOMYOMATOSIS PERITONEALIS DISSEMINATA

Left: Low-power microscopic view of nodules of eosinophilic spindle cells in a background of omental adipose tissue.
Right: Higher-power view shows smooth muscle cells with bland nuclear features and absence of mitotic figures.

WALTHARD RESTS

Walthard rests arise in the serosa of the fallopian tubes and mesovarium, or within the ovarian hilus. They probably derive from foci of epithelium of müllerian origin. They appear as small glistening lesions and may be solid or cystic. Histologically, they are composed of rounded collections of flattened to cuboidal cells that have an appearance resembling transitional epithelium. Rarely, Brenner tumors may arise from this epithelium (49). Walthard rests should not be confused with metastatic deposits.

SPLENOSIS

Splenosis typically occurs as a result of implantation of splenic tissue after traumatic rupture of the spleen or after splenectomy (16,32). Most patients are asymptomatic and the lesions are incidental findings at laparotomy or autopsy. The clinical picture and gross appearance in women may resemble endometriosis or peritoneal carcinomatosis, with nodules ranging from microscopic to several centimeters in diameter (16). Thoracic splenosis has also been described (40).

MELANOSIS

Benign *peritoneal melanosis* is a rarely reported condition characterized by focal or diffuse brown peritoneal pigmentation or dark tumor-like nodules within the pelvis and omentum. Most documented cases have been associated with ovarian mature cystic teratomas, but rare cases have been associated with ovarian serous cystadenoma, enteric duplication cyst, or gastric triplication (12,18,30,38). The lesion consists of collections of pigment-laden macrophages with bland cytologic features. Recent studies have suggested that, in at least some of these cases, the pigment is not actually melanin but rather pigment derived from hemorrhage in the presence of the acid and lipid associated with peptic ulceration (28).

TROPHOBLASTIC IMPLANTS

Trophoblastic implants occur as a postoperative complication of treatment for tubal pregnancy (2,43,45). Implants of villi and trophoblastic tissue occur in the pelvic peritoneum or omentum, and may be a source of postoperative intra-abdominal hemorrhage. Persistent elevation of beta-human chorionic gonadotropin levels may be observed during the postoperative period and is a clue to the diagnosis. These implants are more likely to occur in cases managed by laparoscopy than in those managed by laparotomy.

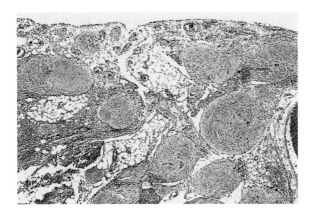 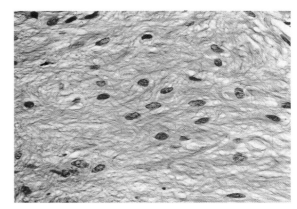

Figure 8-10

GLIOMATOSIS PERITONEI

Left: Low-power magnification shows nests of glial tissue within the peritoneum.
Right: Higher magnification shows glial tissue consisting of spindle cells in a fibrillary matrix. (Courtesy of Dr. S. Robboy, Durham, NC, Robboy Associates, LLC.)

GLIOMATOSIS PERITONEI

Benign mature glial implants have been reported to occur on the peritoneal surfaces (fig. 8-10) (15,26,37,51). This condition, known as *gliomatosis peritonei,* is a rare complication of solid or immature ovarian teratoma, usually in young girls. Rare cases have also been reported in association with ventriculoperitoneal shunts (26). Coexistence with endometriosis has been described (37). Recent studies suggest that the lesion arises from metaplasia of peritoneal tissues rather than from true neoplastic implants (15). The lesion may persist or eventually disappear. Rare malignant transformation has been reported (51).

INFLAMMATORY MYOFIBROBLASTIC TUMOR (INFLAMMATORY PSEUDOTUMOR)

Inflammatory myofibroblastic tumors have been identified in multiple sites of the body, including the abdomen (9,11,44). Patients are typically young and present with a mesenteric mass, fever, weight loss, anemia, thrombocytosis, or polyclonal hypergammaglobulinemia. The disease has a benign clinical course, with surgical resection usually curative but with local recurrence in some cases (11,44). Distant metastases have not been reported (9).

Macroscopically, these tumors have a firm, white appearance, with infiltrative borders and focal myxoid change. There are histologic pat-

terns: 1) myxoid, vascular, and inflammatory areas resembling nodular fasciitis; 2) compact spindle cells with intermingled inflammatory cells resembling fibrous histiocytoma; and 3) dense plate-like collagen resembling desmoid or scar. Immunohistochemical studies show that the spindle cells stain for vimentin, smooth muscle actin, muscle-specific actin, and cytokeratins, consistent with a myofibroblastic origin (9).

OMENTAL-MESENTERIC MYXOID HAMARTOMA

Omental-mesenteric myxoid hamartoma is a novel entity first described by Gonzalez-Crussi et al (19). The lesion consists of multiple omental and mesenteric nodules, which may be confused with metastatic implants. Microscopically, there is a richly vascularized, myxoid stroma with plump mesenchymal cells. Follow-up has demonstrated a benign clinical course without recurrence (19,20). This may represent a variant of inflammatory myofibroblastic tumor (9).

PSEUDOMYXOMA PERITONEI

Diffuse involvement of the peritoneum by mucinous implants is called *pseudomyxoma peritonei*. These lesions typically arise from a primary mucinous cystadenoma or cystadenocarcinoma of the appendix (fig. 8-11), but rarely can arise from a mucinous component of ovarian mature cystic teratoma (46,47). The implants

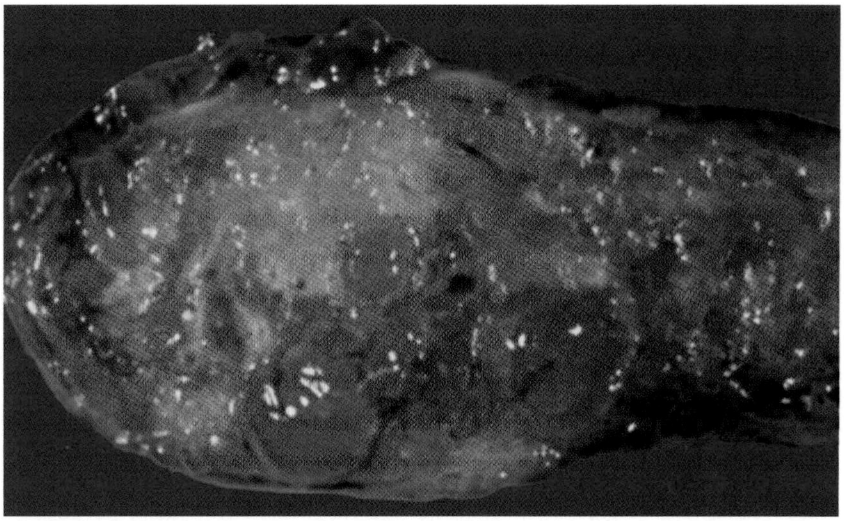

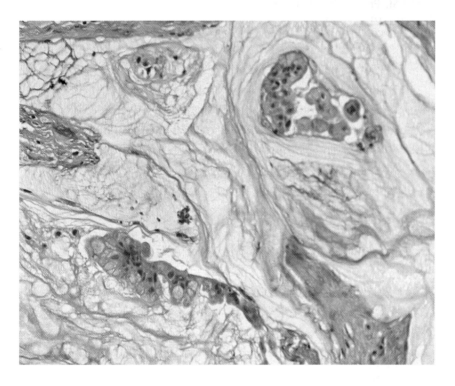

Figure 8-11

PSEUDOMYXOMA PERITONEI

Multiple gelatinous deposits throughout the abdomen give this lesion the name "jelly belly."

Figure 8-12

PSEUDOMYXOMA PERITONEI

There is abundant mucin in association with scattered clusters of tumor cells.

grossly have a gelatinous appearance, giving rise to the term "jelly belly." Pseudomyxoma peritonei may be separated into two diagnostic categories: those cases composed primarily of extracellular mucin are designated *disseminated peritoneal adenomucinosis*, whereas lesions characterized by abundant atypical mucinous epithelium are classified as *peritoneal mucinous carcinomatosis* (fig. 8-12). This classification has been shown to have prognostic significance (47). The tumor cells are positive for cytokeratin 20 and negative for cytokeratin 7. Surgery is the mainstay of treatment, with the role of intraperitoneal and systemic chemotherapy poorly defined (23).

REFERENCES

1. Adams AL, Castro CY, Singh SP, Moran CA. Pleural amyloidosis mimicking mesothelioma: a clinicopathologic study of two cases. Ann Diagn Pathol 2001;5:229–32.
2. Ben-Arie A, Goldschmit R, Dgani R, et al. Trophoblastic peritoneal implants after laparoscopic treatment of ectopic pregnancy. Eur J Obstet Gynecol Reprod Biol 2001;96:113–5.
3. Bontemps F, Tillie-Leblond I, Coppin MC, et al. Pleural amyloidosis: thoracoscopic aspects. Eur Respir J 1995;8:1025–7.
4. Burmeister RE, Fechner RE, Franklin RR. Endosalpingiosis of the peritoneum. Obstet Gynecol 1969;34:310–8.
5. Butnor KJ, Burchette JL, Robboy SJ. Progesterone receptor activity in leiomyomatosis peritonealis disseminata. Int J Gynecol Pathol 1999;18:259–64.
6. Chan JK, Loo KT, Yau BK, Lam SY. Nodular histiocytic/mesothelial hyperplasia: a lesion potentially mistaken for a neoplasm in transbronchial biopsy. Am J Surg Pathol 1997;21:658–63.
7. Churg A. Nonneoplastic disease caused by asbestos. In: Churg A, Green FH, eds. Pathology of occupational lung disease, 2nd ed. Baltimore: Williams & Wilkins; 1998:277–338.
8. Clement PB, Young RH. Florid cystic endosalpingiosis with tumor-like manifestations: a report of four cases including the first reported cases of transmural endosalpingiosis of the uterus. Am J Surg Pathol 1999;23:166–75.
9. Coffin CM, Watterson J, Priest JR, Dehner LP. Extrapulmonary inflammatory myofibroblastic tumor (inflammatory pseudotumor). A clinicopathologic and immunohistochemical study of 84 cases. Am J Surg Pathol 1995;19:859–72.
10. Cooper P. Mixed mesodermal tumor and clear cell carcinoma arising in ovarian endometriosis. Cancer 1978;42:2827–31.
11. Day DL, Sane S, Dehner LP. Inflammatory pseudotumor of the mesentery and small intestine. Pediatr Radiol 1986;16:210–5.
12. De la Torre Mondragon L, Daza DC, Bustamante AP, Fascinetto GV. Gastric triplication and peritoneal melanosis. J Pediatr Surg 1997;32:1773–5.
13. Deering S, Miller B, Kopelman JN, Reed M. Recurrent leiomyomatosis peritonealis disseminata exacerbated by in vitro fertilization. Am J Obstet Gynecol 2000;182:725–6.
14. Fadare O, Bifulco C, Carter D, Parkash V. Cartilaginous differentiation in peritoneal tissues: a report of two cases and a review of the literature. Mod Pathol 2002;15:777–80.
15. Ferguson AW, Katabuchi H, Ronnett BM, Cho KR. Glial implants in gliomatosis peritonei arise from normal tissue, not from the associated teratoma. Am J Pathol 2001;159:51–5.
16. Fleming CR, Dickson ER, Harrison EG Jr. Splenosis: autotransplantation of splenic tissue. Am J Med 1976;61:414–9.
17. Fox H, Buckley CH. Current concepts of endometriosis. Clin Obstet Gynaecol 1984;11:279–87.
18. Fukushima M, Sharpe L, Okagaki T. Peritoneal melanosis secondary to benign dermoid cyst of the ovary: a case report with ultrastructural study. Int J Gynecol Pathol 1984;2:403–9.
19. Gonzalez-Crussi F, deMello DE, Sotelo-Avila C. Omental-mesenteric myxoid hamartomas. Infantile lesions simulating malignant tumors. Am J Surg Pathol 1983;7:567–78.
20. Gonzalez-Crussi F, Sotelo-Avila C, deMello DE. Primary peritoneal, omental, and mesenteric tumors in childhood. Sem Diagn Pathol 1986;3:122–37.
21. Griffith LM, Carcangiu M. Sex cord tumor with annular tubules associated with endometriosis of the fallopian tube. Am J Clin Pathol 1991;96:259–62.
22. Hallman KB, Nuhhas WA, Connelly PJ. Endosalpingiosis as a source of psammoma bodies in a Papanicolaou smear. A case report. J Reprod Med 1991;36:675–8.
23. Harshen R, Jyothirmayi R, Mithal N. Pseudomyxoma peritonei. Clin Oncol (R Coll Radiol) 2003;15:73–7.
24. Heaps JM, Nieberg RK, Berek JS. Malignant neoplasms arising in endometriosis. Obstet Gynecol 1990;75:1023–8.
25. Heinig J, Gottschalk I, Cirkel U, Diallo R. Endosalpingiosis—an underestimated cause of chronic pelvic pain or an accidental finding? A retrospective study of 16 cases. Eur J Obstet Gynecol Reprod Biol 2002;103:75–8.
26. Hill DA, Dehner LP, White FV, Langer JC. Gliomatosis peritonei as a complication of a ventriculoperitoneal shunt: case report and review of the literature. J Pediatr Surg 2000;35:497–9.
27. Hillerdal G. Rounded atelectasis. Clinical experience with 74 patients. Chest 1989;95:836–41.
28. Jaworski RC. Peritoneal "melanosis." Int J Gynecol Pathol 2003;22:104.
29. Kavuru MS, Adamo JP, Ahmad M, Mehta AC, Gephardt GN. Amyloidosis and pleural disease. Chest 1990;98:20–3.

143

30. Kim NR, Suh YL, Song SY, Ahn G. Peritoneal melanosis combined with serous cystadenoma of the ovary: a case report and literature review. Pathol Int 2002;52:724–9.

31. Knapp MJ, Roggli VL, Kim J, Moore JO, Shelburne JD. Pleural amyloidosis. Arch Pathol Lab Med 1988;112:57–60.

32. Kumar RJ, Borzi PA. Splenosis in a port site after laparoscopic splenectomy. Surg Endosc 2001; 15:413–4.

33. Kuo T, London SN, Dinh TV. Endometriosis occurring in leiomyomatosis peritonealis disseminata: ultrastructural study and histogenetic consideration. Am J Surg Pathol 1980;4:197–204.

34. Lynch DA, Gamsu G, Ray CS, Aberle DR. Asbestos related focal lung masses: manifestations on conventional and high resolution CT scans. Radiology 1988;169:603–7.

35. Mahoney AD, Waisman J, Zeldis LJ. Adenomyoma: a precursor of extrauterine Mullerian adenocarcinoma? Arch Pathol Lab Med 1977;101:579–84.

36. Malpica A, Deavers MT, Shahab I. Gross deciduosis peritonei obstructing labor: a case report and review of the literature. Int J Gynecol Pathol 2002;21:273–5.

37. Muller AM, Sondgen D, Strunz R, Muller KM. Gliomatosis peritonei: a report of two cases and review of the literature. Eur J Obstet Gynecol Reprod Biol 2002;100:213–22.

38. Nada R, Vaiphei K, Rao KL. Enteric duplication cyst associated with melanosis peritonei. Indian J Gastroenterol 2000;19:140–1.

39. Nascimento AG, Keeny GL, Fletcher CD. Deciduoid peritoneal mesothelioma. An unusual phenotype affecting young females. Am J Surg Pathol 1994;18:439–45.

40. O'Connor JV, Brown CC, Thomas JK, Williams J, Wallsh E. Thoracic splenosis. Ann Thorac Surg 1998;66:552–3.

41. Ordonez NG, Ro JY, Ayala AG. Lesions described as nodular mesothelial hyperplasia are primarily composed of histiocytes. Am J Surg Pathol 1998;22:285–92.

42. Oury TD. Benign asbestos-related pleural diseases. In: Roggli VL, Oury TD, Sporn TA, eds. Pathology of asbestos-associated diseases, 2nd ed. New York: Springer; 2004:169–92.

43. Pal L, Parkash V, Rutherford TJ. Omental trophoblastic implants and hemoperitoneum after laparoscopic salpingostomy for ectopic pregnancy. A case report. J Reprod Med 2003;48:57–9.

44. Pettinato G, Manivel JC, De Rosa N, Dehner LP. Inflammatory myofibroblastic tumor (plasma cell granuloma): clinicopathologic study of 20 cases with immunohistochemical and ultrastructural observations. Am J Clin Pathol 1990; 94:537–46.

45. Reich H, DeCaprio J, McGlynn F, Wilkie WL, Longo S. Peritoneal trophoblastic tissue implants after laparoscopic treatment of tubal ectopic pregnancy. Fertil Steril 1989;52:337–9.

46. Ronnett BM, Seidman JD. Mucinous tumors arising in ovarian mature cystic teratomas: relationship to the clinical syndrome of pseudomyxoma peritonei. Am J Surg Pathol 2003;27: 650–7.

47. Ronnett BM, Zahn CM, Kurman RJ, Kass ME, Sugarbaker PH, Shmookler BM. Disseminated peritoneal adenomucinosis and peritoneal mucinous carcinomatosis. A clinicopathologic analysis of 109 cases with emphasis on distinguishing pathologic features, site of origin, prognosis, and relationship to "pseudomyxoma peritonei." Am J Surg Pathol 1995;19:1390–408.

48. Rosai J, Dehner LP. Nodular mesothelial hyperplasia in hernia sacs: a benign reactive condition simulating a neoplastic process. Cancer 1975;35:165–75.

49. Roth LM. The Brenner tumor and the Walthard cell nest. An electron microscopic study. Lab Invest 1974;31:15–23.

50. Rutgers JL, Young RH, Scully RE. Ovarian yolk sac tumor arising from an endometrioid carcinoma. Hum Pathol 1987;18:1296–9.

51. Truong LD, Jurco S, McGavran MH. Gliomatosis peritonei. Report of two cases and review of the literature. Am J Surg Pathol 1982;6:443–9.

52. Veinot JP, Tazelaar HD, Edwards WD, Colby TV. Mesothelial/monocytic incidental cardiac excrescences: cardiac MICE. Mod Pathol 1994;7:9–16.

53. Wilson JD, Montague CJ, Salcuni P, Bordi C, Rosai J. Heterotopic mesenteric ossification ('intraabdominal myositis ossificans'): report of five cases. Am J Surg Pathol 1999;23:1464–70.

54. Zaytsev P, Taxy JB. Pregnancy-associated ectopic decidua. Am J Surg Pathol 1987;11:526–30.

Index*

---

*Numbers in boldface indicate table and figure pages.